3/00

THE REAL SCIENCE BEHIND

T H E (X) — F I L E S

MICROBES, METEORITES, AND MUTANTS

Anne Simon, Ph.D.

Simon & Schuster

SIMON & SCHUSTER
Rockefeller Center
1230 Avenue of the Americas
New York, NY 10020

SIMON & SCHUSTER and colophon are registered
trademarks of Simon & Schuster, Inc.

Designed by Ruth Lee

Manufactured in the United States of America

10 9 8 7 6 5 4 3 2 1

Library of Congress Cataloging-in-Publication Data

Simon, Anne Elizabeth, 1956–
The real science behind the X-files : microbes,
meteorites, and mutants / Anne Simon.
p.   cm.
1. Life sciences. 2. X-files (Television program).
3. Genetics. I. Title.
QH309.S554 1999
570—dc21   99-16531
                                CIP

ISBN 0-684-85617-4

To MAYO SIMON,

for passing down the writing genes,

and SONDRA SIMON,

for being the world's greatest mom.

Contents

THE REAL SCIENCE BEHIND

T H E (X) – F I L E S

Foreword

MULDER

I have plenty of theories. What has me stumped is why Bureau
policy is to label these cases as unexplained phenomena and
ignore them.
                    (to the point)
Do you believe in the existence of extraterrestrials?

SCULLY

I've never given it much thought.

MULDER

As a scientist.

SCULLY

Logically, I'd have to say no. Given the distances needed to
travel from the far reaches of space, the energy requirements would exceed—

MULDER

—Conventional wisdom. That girl in Oregon—she's the fourth
member of her graduating class to die under mysterious circumstances. When convention and science offer no answers,
might we not consider the fantastic as a plausibility?

"The X-Files Pilot Episode, 1992"

I'm often asked a question by strangers, by reporters, by TV executives, and by fans of *The X-Files:* Where do you get your ideas? It's an obvious question given the kind of show *The X-Files* is, but I never have a good comeback and usually try to make a joke. It seems counter to the creative process to give a straight answer. Where do they think the ideas come from? They come from our imaginations—mine and the other writers' on *The X-Files*. Likewise the characters of Mulder and Scully, who'd come to me bidden and unbidden in the summer of 1992 when I was creating the show. The process is mysterious, or as the scientific Scully would tell you, simply an unexplained phenomenon. The truth is out there.

Actually, the truth is, more often than not, the ideas which become *X-Files* stories are rooted in hard science, and even when they are not generated as such, they're built on a foundation of scientific convention. The point of view of the series is essentially Agent Scully's, the scientific counterpoint to Agent Mulder's belief in the supernatural. The rational versus the irrational. Her sober approach is the skeptical counterintuitive to Mulder's postmodern fanaticism. It's her science (she's a medical doctor) on which the science fiction depends. All the forces, creatures, acts, and apparitions that Mulder may throw at her—six years' worth at this writing—she believes science can explain. And if not now, eventually. Her faith in the empirical process is equal to Mulder's in the fantastic. (Even more so, if you consider the "I Want to Believe" poster that hangs on Agent Mulder's office wall; he is working through his convictions, while Agent Scully has her feet planted firmly on the alien-uninhabited ground.)

The problem is, Agent Scully is rarely, if ever, right. Her science is unequal to the wonders of the universe, or at least to the wonders of Mulder's multitude of FBI case files. Prove it, she's asked rhetorically each week, and each week she can't quite. It's not that she's wrong, but by necessity she is left without any good explanation. She and her methods are inadequate and can't ever seem to wipe the smile off Agent Mulder's face at the end of each episode.

And it is a problem, or so it's been pointed out to me, most demonstrably by the Committee for Scientific Investigation into Claims of the Paranormal (CSICOP) who invited me to speak to several hundred of its members (Nobel winners among them) in Buffalo, New York, several summers ago. They had me for lunch, as it were, where I looked around and saw few such smiles as Mulder's in the big university meeting hall. It felt as if I were standing before an army of Agent Scullys who were branding me a prime time purveyor of "pseudoscience." As if I alone were responsible for all the loopy, looney trends in angels and aliens, in superstition, and even in fundamentalism. I was threatening to destroy yet another generation of minds by feeding them more bogus claptrap. Carl Sagan, one of CSICOP's most prominent scientists, had just published *The Demon-Haunted World*, which took to task people just like me.

I was arguably guilty to some degree, the popularity of the show irrefutable evidence of this. Agent Mulder does believe in extraterrestrial life and in the government conspiracy to keep its existence a secret from the world. He does believe in vampires, ghosts, tulpas, revenants, PK, telekinesis, reincarnation, voodoo, astral projection, eidolons, zombies, werewolves, and sewer-dwelling human flukeworms from Chernobyl. (To be fair, what he really believes in is the "extreme possibility" of these things. If he believed in them absolutely, why bother investigating them?!) How could I defend myself, and why even try? To make matters worse, my formal science education ended somewhere around college graduation, and I hadn't been such a great student. I was going to meet the enemy, a large number of whom were college professors, and I was unarmed. From where did a television producer get such audacity? (Some would call that a tautology.)

To this question I did have a good answer, in the name of Anne Simon. Like my accusers, she is a college professor and a skeptic and, like Agent Scully, has a trust and faith in the scientific process. She is not just a teacher but a researcher. Her work with plant viruses has made her part of an international community of scientists pushing the limits of practical and theoretical understanding of plant genetics. I'd met Anne through her mother and father, who are friends of my wife, and near the end of the first season of *The X-Files* I'd learned with pleasant surprise she was a fan of the show. This was at the same time I was working to fully develop the show's "mythology," the alien conspiracy Agent Mulder believed in, so I'd given Anne a call. I told her I was uninterested in aliens per se, in the "literature" and its devotees, and also in the spaceships and looking to the skies (we'd seen so much of this, there hardly seemed anywhere new to go, particularly given TV budgets). I wanted to take a scientific approach to the subject, both psychological and genetic, and to build my extended story line on accepted theories and fact. (This was my intention from the start. I would tell the writers and directors ad nauseam that the show was only as scary as it seemed believable, and only as believable as it seemed real or plausible.) My idea for the first-season series finale was an experiment that involved genetic material Agent Mulder had stumbled on, which Agent Scully would learn was undeniably extraterrestrial QED. From Anne I got the building blocks of the mystery, literally: genes, chromosomes, proteins, nucleotides. I also got a careful script reader who would call me on inaccuracies in the science, and, through our long conversations, I even got from her the title of the episode: "The Erlenmeyer Flask." (The episode would go on to be nominated for an Edgar Award by the Mystery Writers of America.)

Why would Anne Simon throw in with a heretic like me? Because she'd come to science and to being a scientist through her love of science fiction. And this is what I told the CSICOP members over lunch. Anne has been and continues to be a voracious reader of science fiction novels, both the futuristic and the more speculative works closer to the stories that we tell on *The X-Files*. She, like many scientists I've met, has great enthusiasm for her work, and approaches it with imagination and appreciation for ideas. Which is what *X-Files* writers do, actually. Creative speculation seems to be the force that drives science forward, rather than the limiting conventions of accepted truth. I'm reminded of Dr. Tom Kaufman, a scientist Anne suggested I contact at Indiana University, with whom I spent a wonderful day learning about his genetic experiments on the *Drosophila* fly. His work became the basis for a Frankenstein story I wrote called, "The Postmodern Prometheus," about an amateur scientist who creates a son by manipulating human genes like Dr. Kaufman does his flies'. This is something we may very well see applied in our lifetimes. (I also told the committee about another professor with a love of science fiction. Someone much closer to me, my younger brother Craig Carter, who teaches material science at MIT.)

If Anne Simon was affected by her early exposure to ideas, might not *The X-Files* inspire future scientists? I believe it will, and could very well continue to for years to come. It's fair to say no other long-running dramatic television show has used such a wide (and strange!) variety of science fact in its storytelling: medicine, genetics, cryonics, cosmology, quantum physics, dendrochronology, to name but a few disciplines. And this in a medium criticized for pandering to the masses. And might the science fiction of *The X-Files* even be regarded someday as science convention? Both a sheep named Dolly and a rock from Mars have made news during the show's run. (Meeting Stephen Hawking some years after my CSICOP lunch, I asked him his feelings about science fiction or pseudoscience. His reply was that science and science fiction had something to give each other, and that "science fiction is no more pseudo than cosmology.")

What I believe Anne Simon ultimately appreciates, and what I asserted to my hosts that afternoon in Buffalo, is that stories and storytelling are essential to life. As Nietzsche and Ibsen knew, life requires life-supporting myths and metaphors. Or even illusions. Freud said myths are public dreams and dreams private myths, both essential to the psyche. Or the soul. Science demystifies the world. It's meant to reassure us, as Georges Braque set forth, whereas the purpose of art is to disturb. The relationship between the two is also essential and should be fostered and celebrated rather than rooted out. Or, as I told the CSICOP folks, science tells you your cell phone might give you a brain tumor, whereas art allows it might be how your long dead Uncle Harry reaches you from the Great Beyond. For which I got a polite round of applause, and after signing a few autographs, was bid adieu. I've never heard from them since.

—Chris Carter, creator of
The X-Files, March 1999

Introduction

In September 1993 I read a description in *TV Guide* for a new show called *The X-Files*. As a fan of science fiction, I found the synopsis to be intriguing—the adventures of two FBI agents who investigate cases of a paranormal nature. My expectations were not particularly high given the channel the show was on. The still fledgling Fox network was better known for urban situation comedies and bubblegum-chewing narcs than serious science fiction.

Was I surprised. The first few minutes of the show were mesmerizing: spooky music, brooding actors, and realistic dialogue added up to an intelligence and quality unusual for any network. When the opening credits rolled and I saw the name of the show's creator, Chris Carter, I thought, could this possibly be the Chris Carter whom I knew? The writer-surfer who married my mother's good friend, scriptwriter Dori Pierson? Last I had heard, Chris and Dori were working for Disney's Buena Vista Studios. Still, knowing how often my scriptwriter father moved from one job to the next meant that anything was possible.

By the end of the first episode, I was hooked. Not only was *The X-Files* great science fiction, but as a bonus, the lead character was a woman doctor with a "background in the hard sciences." A scientist as a main character in a serious television show. And one who was not a nerdy, bow-tied, absentminded, congenial fool or a sinister madman developing a formula to destroy the world. The refreshing Dana Scully was actually being portrayed as a realistic scientist. Attractive (I did say this was realistic), intelligent, and dedicated to her work, Scully was a character I, a woman scientist, could relate to.

Without the efforts of my eighth-grade biology teacher, I might never have become a scientist. Mrs. Webb had inexplicably placed me in a ninth-grade science class as my first elective course at Paul Revere Junior High School. This gesture, poorly appreciated at the time, led to classes at the Los Angeles Museum of Natural History followed by enrollment at the University of California at San Diego. My interests turned from marine biology, the initial major of many from the Pacific Palisades fun-in-the-sun crowd, to genetics after I took some stimulating college courses. In my senior year, one of my professors suggested Indiana University for graduate school. Four years later, I received a Ph.D. in genetics for solving the mystery of why some animal cells have mutation rates far beyond the norm. By the way, aliens were not involved.

After finishing graduate school, I needed to remain at Indiana University while I waited for my husband to get his Ph.D. In order to stay at the same university, I was advised—strongly—to switch fields; otherwise I put future academic positions at risk. (To clarify the logic of this advice to nonacademicians, switching fields shows the proper adventurous spirit thought necessary for a budding scientist, which counters the more timid impression of not wanting to transfer away from one's training environment.) As a result, I decided to study how plant embryos develop. Not thoroughly enamored of the subject and eager to revisit the Pacific Ocean, I headed back to San Diego and began studying viruses (by this time, relatives were wondering if I was ever going to get a "real" job). Viruses and I agreed with each other, and I continued in the field of virology after getting that real job as an assistant professor at the University of Massachusetts at Amherst.

Five years after settling down to a life of research and teaching, and midway through *The X-Files'* first season, I received a phone call from my mother. Did I know that Dori's husband, Chris Carter, had a new series called *The X-Files*? (So it *was* Chris's show.) Chris, knowing that I was a scientist, had asked my mother if I would mind discussing some science questions for one of his scripts. She told me that she had given him my phone number. Moms do know best.

Chris called the following day. He was delighted to hear that I was a fan of the show, particularly since the ratings were not quite what they are today. Chris described his idea for an episode in which a scientist suffering from cancer finds some alien tissue that might be a cure. The scientist is planning to test the tissue for harmful effects on children at a large state-run institution (paralleling a news story several years ago about scientific experiments performed on institutionalized children without their consent). I told Chris that a scientist dying of cancer would probably subject himself to the experiment given the long history of scientists experimenting on themselves. Chris liked that idea and went on to ask: If someone handed a microbiologist an unclassified microbe, how would it be studied? I described three steps: grow more of the organism by culturing it in an Erlenmeyer flask; visualize it under a microscope; and examine its genetic material, its DNA. After going into depth on all three of these procedures, Chris needed an answer to his most important question: What experimental result would instantly suggest an extraterrestrial origin for the organism? Chris was waiting, so I quickly thought up a little science with a science fiction twist—some results that would make me reach for the phone if I ever encountered such a strange microbe.[1]

A few weeks after our conversation, the finished script for "The Erlenmeyer Flask" arrived by express mail. Since I was used to reading my father's science fiction scripts (*Marooned, Phase IV, Futureworld, Man from Atlantis*), I could vividly picture the fantastic episode that would emerge on the screen. There were, however, a few scientific inaccuracies in the description of one scientist's area of expertise—the human genome project—and in the conversation between Scully and scientist Anne Carpenter. But these problems were easily corrected, mostly with the change of a single word or phrase. When I watched "The Erlenmeyer Flask" on TV, I followed along with my copy of the script. To my delight, Chris had used every suggestion. Soon afterward, he sent me one of the first *X-Files* T-shirts emblazoned with the phrase "The Truth Is Out There." I remember thinking that he was crazy to have T-shirts made up with the logo from a TV show and expect people to buy them. In retrospect, this thought

[1]Explained in detail in Chapter 2.

was ample evidence that an aptitude for science doesn't necessarily translate into a head for business.

As *X-Files* episodes continued to air, the central nature of science and Scully's role in the series became evident. Scully provides realistic scientific interpretations behind the decidedly odd events. Scully is the quintessential scientist. She gathers information and bases her hypotheses on that evidence. She also keeps partner Fox Mulder from rushing to unsupported conclusions. Mulder, searching for the answers behind his sister's childhood abduction, readily formulates the most outlandish explanations for what are, granted, rather unusual happenings. If Mulder finds that a person has been buried in mud in a standing position and a corpse is removed from a coffin, then it must be the trees in an orchard that are killing people and raiding graves, guided by the psychic emanations from a deranged woman possessed by the spirit of her abusive father. Scully, adhering to the scientifically sound precept that the simplest explanation is likely to be correct, tries to convince Mulder that the dead root system of the diseased trees combined with substantial rainfall created the muddy sinkholes that swallowed up the townspeople.

And therein lies the controversy. Controversy that reaches all the way to the editorial pages of one of the world's most eminent scientific periodicals, the British journal *Nature*. For as *X-Files* fans know, Scully is usually wrong. The trees *are* being guided by the brain waves of a psychotic woman. The concept that science cannot explain all "unnatural" occurrences and the believable nature of the show's science fiction scenarios leave *The X-Files* open to critics who claim that it is "antiscience." What the critics of the show have lost sight of is that *The X-Files* is science fiction. If Scully's mundane explanations were correct more often, it is doubtful that the series would have lasted into its second season, let alone achieve its current cult status.

Those who see *The X-Files* as promoting pseudoscience (and therefore antiscience) are missing the point. Viewers, especially high school and college students, who make up the core of the fan base, are seeing scientists portrayed in a favorable light, perhaps for the first time. Few outside of the scientific world are personally acquainted with a scientist and therefore have only inaccurate fictional characterizations (usually unfavorable) with which to base their feelings. The portrayal of scientists in most television and feature films as remote, emotionless, obsessed individuals does little to attract bright young minds to the fascinating world of science.

Granted, the scientists in *The X-Files* don't have all the answers and their results are frequently open to many possible interpretations. But this is precisely what is faced by research scientists on a daily basis. The goal of scientists is to solve mysteries: How do viruses reproduce? Why does a particular virus infect some organisms and not others? Why are some viruses deadly while others go unnoticed? These are just some of the puzzles that I face daily in my own research. And just like the adventures of Mulder and Scully, answers lead to more puzzles, dead ends loom around every corner, and a mind open to remote possibilities is the key ingredient behind finding the truth.

The characters on *The X-Files*—some of them, anyway—are pretty accurate portraits of contemporary working scientists. Dana Scully doesn't claim a knowledge of botany through nuclear physics (like a certain stranded castaway). Rather, she uses her medical degree to perform autopsies and her knowledge of genetics and biochemistry to conduct experiments. As with any scientist, when the investigation leads to areas outside her expertise, she consults with other experts. In a refreshing departure from the norm, scientists on *The X-Files* are more likely to be aiding the investigations than perpetrating the crimes. While many bizarre and completely fictional creatures populate *X-Files* episodes, the scientific investigations of these creatures are based in reality. The proper experiments are conducted; the correct microscopes are used; evidence is gathered and conclusions are based on that evidence. To achieve such accuracy on the show requires an attention to detail and extra effort from the writers that fans can see and appreciate— and many of these fans are scientists.

I am often asked why I care so much about scientific accuracy on a science fiction show. The answer is simple. As a scientist, it usually isn't possible to watch science in movies or TV without wincing. The microscopes are wrong; the language is wrong; cures of viral infections are instantaneous; organisms are described as being part bacteria, part virus (analogous to saying part watermelon, part speck of dust). What I find frustrating is that many problems could have been fixed without affecting the plot—if only the writers had spent a few minutes consulting with a working scientist.

My association with *The X-Files* has involved a number of scripts (all written by Chris Carter, a true creative genius), including the *X-Files* movie, *Fight the Future*. I feel fortunate to have played a small role in helping to ground *The X-Files* in real science. However, the average viewer of a television show that prides itself on its depiction of realistic and very scary creatures can have difficulty deciphering the line between science and science fiction. The goal of this book is to explain to nonscientists the real science behind *The X-Files*. To use the show as a springboard to examine the many science issues that are blended into plots—hot topics like cloning, aging, genetic engineering, and life on other planets. In an age where science is transforming the food we eat, the information that we process, and the health care we receive, knowledge of basic scientific tenets can no longer be thought of as too complicated, too boring, or confined to the realm of stereotypic white-coated geeks. Besides the mere facts, I also hope to convey the excitement of biological science, which abounds with creatures and mysteries every bit as strange as any appearing on *The X-Files*.

Enjoy the journey.

1

Hidden and Hungry

Introduction

```
MIDDLESEX COUNTY PSYCH HOSPITAL—DAY
ANGLE ON CORNER OF CELL
```

Where, squeezed back behind a series of pipes running vertically up the wall, A GRAYISH HUMAN FORM remains perfectly still. Its skin is mottled and scarred and it appears to be smooth and hairless, coated with a shiny layer of clear slime, like a snail or a slug. Because its face is obscured by the pipes, the only way you can tell it is alive is by an unnatural pulsing in its neck, similar to a bullfrog in this respect.

```
INT. CORRIDOR—CONTINUOUS
RESUME MULDER, SCULLY IN CORRIDOR
```

MULDER
I don't know if you can see, but it has no sex organs. It's genderless.

SCULLY
Platyhelminthes are often hermaphroditic. This is amazing, Mulder. Its vestigial features look parasitic, but it seems to have primate physiology.

(beat)

Where the hell did it come from?

MULDER

(with due irony)

I don't know. But it looks like I'm going to have to tell Skinner that the suspect is a blood sucking worm after all.

—"The Host"

The largest life form on Earth is the six-thousand-ton intercon-
nected quaking aspen in Utah. The smallest is bacteria, about eight
hundred thousandths of an inch across. With living creatures available in virtually every size in between, life is present in seemingly infinite varieties. Unless you're surrounded by a barren desert, walk
outside and look around at the hundreds of plants, insects, birds,
mammals, and fungi in your vicinity. Peer into a drop of pond water
and you'll find a world of microorganisms that when magnified look
every bit as uncanny as the *X-Files'* legendary flukeman. Any spoon-
ful of dirt contains thousands of different bacteria, many of which
have never been identified. At this very moment, in the cubic yard of
air at the tip of your nose, hundreds of thousands of microscopic bacteria, viruses, fungal spores, algae, and pollen grains are floating by.
Over 50 million species share this planet with us, the products of
nearly 4 billion years of evolution.

I probably wouldn't get an argument from even the most devoted *X-Files* fans if I pointed out that flukemen weren't likely to ever
peek out of your toilet. What comes closer to the line between science and science fiction, and is therefore much more disquieting, are
the *X-Files* creatures that don't require a supernatural origin. These
are the organisms hidden in places not normally seen by humans—in
volcanic rocks, within ancient trees, frozen beneath the arctic ice, or
deep in the heart of unexplored rain forests. Maybe it's simple,
everyday fungi, floating invisibly in the air, which emerge from obscurity to become deadly disease agents. When you consider that
only a few percent of the organisms on planet Earth have been identified, the chance of finding new and not always friendly creatures is
not only within the realm of possibility, it is a virtual certainty.

Each day, previously unknown creatures are being discovered, and not just in exotic locations far from civilization. In my own baili- wick of Massachusetts, naturalists have been examining fields, forests, lakes, and rivers for almost three hundred years—plenty of time to turn over every rock, dig through every swamp, and traverse every forest. But all it takes is a dip in the Connecticut River—the first river in the United States to be navigated by European settlers— to realize the fallacy of such a statement. The Connecticut River is a beautiful, pastoral body of water that slowly meanders four hundred miles from the Canadian border through the fertile valleys of West- ern Massachusetts before finishing its journey at the Long Island Sound. Recent underwater explorations by my University of Massa- chusetts colleague Dr. Edward Klekowski revealed the remains of a previously unknown Ice Age lake hidden below the rippling surface. The proglacial Lake Hitchcock left an imprint of clay sediments in the Connecticut River that are now covered with enormous numbers of larvae of a fly not known before to exist. Other strange creatures in- habit deeper regions of the river. A crack in the Earth some 200 mil- lion years ago caused the base of the river to drop 130 feet below the surface in some locations. In the inky depths lies a world only re- cently explored—a world of giant sponges and mosslike animals that form colonies eerily resembling plants. While these creatures may not spark the imagination like Scotland's Loch Ness Monster or the *X-Files'* own Big Blue Serpent of Heuvelmans Lake, they are real-life equivalents—organisms imagined but never before seen.

Twenty years ago, scientists would have limited the regions where life abounds to the surface of the land and ocean depths where light is not completely absent. After all, living creatures need energy and the primary source of energy is the sun. Scientists figured out long ago that green plants and some microorganisms use the sun's energy directly in the process called photosynthesis—the com- bining of carbon dioxide and water to form sugars. These sugars then serve as a secondary source of energy for the plants and a primary source of energy for the rest of us. Only on land and in the ocean were oxygen and minerals thought to be sufficiently abundant to form the molecules imperative for life. Underground worlds filled with real-life equivalents of Jules Verne's Cretaceous creatures seemed biologically impossible.

These notions disappeared in 1979 when an astonishing article appeared in the journal *Science*. The most remote region thought to exist at the time, the bottom of deep ocean trenches, had been considered by scientists to be a graveyard of dying crabs and rotting fish. Instead, a world was found teeming with life that did not rely on the sun. Researchers were amazed to discover acres of mollusks and tube worms, some over ten feet long, feeding on bacteria that also don't need the sun. Rather, the bacteria derive energy from the blistering furnace of the Earth's interior. Liquid methane and gaseous hydrogen sulfide that slowly seep through hydrothermal vents contain enough chemical energy for deep-dwelling bacteria to generate other types of cellular energy. Bacteria that live near these holes in the ocean's crust must survive in sizzling temperatures once thought to be unendurable for living creatures. Temperatures reach 235°F near ocean vents, and only the extreme pressure of the depths keeps nearby seawater from boiling. What was once considered the most inhospitable of habitats is now believed to be a warm and homey domicile for wet denizens of the very deep.

Finding organisms that are able to use energy sources other than the sun opened up another intriguing possibility. If life doesn't require the sun, couldn't there be life below our feet? Not the dinosaurs of Verne's lost world, but maybe tiny creatures able to live within the near-solid rock. One of the most astonishing scientific discoveries of the past ten years was finding that the abundance and variety of organisms living on the surface of the Earth and in the water was only the tip of life's iceberg. Hidden beneath our feet, below the ocean floor and the arctic ice caps, in temperatures reaching 235°F and in rocks so dense that water can take centuries to permeate, live communities of microorganisms. And not just a few isolated species. Tens of thousands of different strains of bacteria have already been identified that live as much as 2.5 miles below the continental crust, the very limits that sweltering temperatures permit life. Any lower down and the sizzling heat would cause a cell's biomolecules to be destroyed faster than they can function.

Dr. Thomas Gold of Cornell University has calculated that pound for pound, there may be as much life living below the ground as there is life on the planet's surface. Creatures that live in this hidden underground world have been separated from surface dwellers for hundreds of millions of years. There is no oxygen in most places, so organisms must find a way to survive in its absence. There is little water or food, so creatures must learn to live while thirsty and starving. These seemingly miserable microorganisms are encased in tiny crevasses in rock and need hundreds, maybe even thousands of years to reproduce by dividing their tiny selves in two.

But if you're simply looking for some new species, it isn't necessary to drill miles underground or plunge into deep water. The earth is occupied by an amazing variety of creatures whose universe is the body of another organism. Endosymbiont is what biologists call creatures that live in harmony with their host in a mutually beneficial arrangement. Endopathogens exploit the host for their own gain; by the time an endopathogen is finished with its host, the host is usually dead.

As humans, we like to think that we are the masters of our bodies. But sharing our innards are billions of bacteria, protozoa, viruses, and fungi. These invisible microbes are everywhere—in our mouth, ears, nose, stomach, and skin. Just days after a baby is born, the microbes start pouring in. Cough, and 10 million microbes shoot out into the air. Scrubbing the scalp to remove the millions of bacteria residing in each square inch is only a temporary fix since other little microbes soon move in. There are, by the way, more bacteria in your gut alone than there are human cells in your entire body.

If your perceptions of the world are similar to Fox Mulder's, you might wonder if the microbes inhabiting your body are engaged in a sinister conspiracy instigated by evil forces in the government to take over your carcass. Dana Scully, however, would explain that you wouldn't be alive and healthy without your little guests. Most colonizing microbes are endosymbionts—you scratch their tiny backs and they scratch yours. Microbes in your intestines make several vitamins for your well-being. *Escherichia coli* bacteria are like little factories making and exporting vitamin K and some B vitamins. If you aren't getting enough protein in your diet, *Klebsiella* bacteria can provide your cells with some raw chemical materials needed to help you make proteins. Some of the hundred thousand bacteria per square inch of your skin help protect you from less benevolent microbes. One of your surface bacteria, *Propionibacterium*, produces an acid that helps keep harmful bacteria away, such as the typhoid bacteria, *Salmonella typhi*.

But while some microbes are friendly, others are decidedly not. Infectious diseases have been a thorn in the side of man since civilization began. Concentrate enough people in one location and creatures that cause disease, known as pathogens, can spread from one person to another. When trade routes to nearby towns were established, pathogens tagged along with the caravans. Intrepid explorers seeking adventure beyond the ocean's horizon brought little stowaways to new continents—rats, fleas, lice, bacteria, and viruses. The history of man is a history of infectious diseases. Epidemics of plague, smallpox, typhus, cholera, measles, and a host of plant pathogens wiped out armies, caused widespread famine, and decimated cities. The bubonic plague or "black death" in the fourteenth century indiscriminately killed 25 million Europeans. The influenza pandemic in 1918 killed a half percent of the world's population, some 20 million to 25 million people, including 600,000 Americans, most in the prime of their life. It wasn't the ingenuity of FBI agents that finally conquered the plague bacterium in the Middle Ages or the virus responsible for the lethal influenza. People susceptible to the bacteria or virus died. People who were naturally resistant lived. The pathogens died out when they ran out of people to infect.

While pathogens in *The X-Files* tend to come from exotic locations—beneath the arctic ice or within tropical rain forests—many new diseases in the past twenty-five years have emerged from locations much closer to home. In 1976, 182 people attending the American Legion convention in Philadelphia brought back home with them a deadly souvenir from the Bellevue-Stratford Hotel—a tiny bacterium that had not previously made its presence known. Thirty-four men died of what is now called Legionnaire's disease. The following year, the Centers for Disease Control (CDC) isolated the bacteria responsible for the disease and named it *Legionella*. Since then, dozens of outbreaks of *Legionella* have been reported from San Francisco to Connecticut, with hundreds of other outbreaks around the world.

As anyone with access to a newspaper knows, *E. coli* serotype O157:H7 has become a notorious food contaminant. This strain of *E. coli* was unheard of before 1982. Since then, there have been sixty major outbreaks in the United States. The most severe occurrence was in 1993, when more than seven hundred people in four states became infected by eating undercooked hamburger at a fast food restaurant. Each year in the United States alone, eating food contaminated with dangerous bacteria sickens 6 million to 80 million people, causes 9,000 deaths, and costs an estimated 5 billion dollars. Eating beef isn't the only way to come in contact with this lethal new bacterium. *E. coli* has been found in such unlikely foods as unpasteurized apple juice, salad vegetables, yogurt, and drinking water.

New disease agents are also lurking in forests, lakes, and reservoirs. Lyme disease, caused by bacteria-infected ticks, was not a problem before 1982. Today, it's a common concern when walking in forested areas in the Northeast. Lyme disease emerged from its hiding place within forests when more people decided to live in houses surrounded by trees. *Cryptosporidium,* a protozoan parasite that contaminates drinking water and infects the digestive tract, emerged in 1976. In 1993, the drinking water in Milwaukee, Wisconsin, became contaminated with *Cryptosporidium,* causing more than 400,000 people to become sick and 100 to die. As many as 7 percent of all diarrhea cases in the United States are now attributed to this parasite.

With modern medicine and improvements in technology, why are the numbers of deaths due to infectious diseases climbing higher every year? People are living longer and healthier lives, yet more people will die of tuberculosis this year than in any year in history. Mulder would probably attribute the large numbers of disease outbreaks to the increasing number of reported UFO sightings since the 1940s. However, to find the real reasons for the increase in infectious diseases, one need look no further than, well, modern medicine and improvements in technology.

So why is modern medicine making us sick? Medical technology, like organ transplantation and cancer therapy, means a longer life for people with chronic diseases but also more people with compromised immune systems. This growing segment of the population is most vulnerable to diseases. People with weakened immune systems become reservoirs for high levels of disease agents that can be transmitted to other people. As more people have weakened immune systems, microbes that were once thought of as harmless are now viewed as significant problems. *Cryptosporidium,* for example, is a previously innocuous parasite that now causes between 10 and 20 percent of the intestinal infections in AIDS patients. More ominously, *Cryptosporidium* is now able to infect healthy hosts.

Modern medicine also prolongs the life of people in general. The number of people over the age of seventy-four is 400 percent higher in 1995 than it was in 1950. People who are older are at greater risk for a variety of illnesses. One possible reason why older people are more susceptible to invasion by pathogens is that their stomachs are no longer as acidic, reducing the efficacy of a major body defense against disease agents that are eaten.

Healthier eating habits are also paradoxically contributing to new outbreaks of disease. Those nutritious raw vegetables and fruits can be contaminated with microorganisms while they are being grown, harvested, or delivered to neighborhood supermarkets or restaurants. Cooking vegetables does more than cause vitamins to leach out; it kills many contaminating organisms hidden on the surface or inside foods. Between 1990 and 1997, disease outbreaks have been associated with such healthy pastimes as eating sliced cantaloupe, green onions, unpasteurized cider, freshly squeezed orange juice, lettuce, raspberries, alfalfa sprouts, sliced tomatoes, and frozen strawberries. Our modern lifestyles also include many more visits to restaurants, where 80 percent of reported food-related outbreaks take place.

Modern technology is also a major contributor to increases in harmful microorganisms. Modern food technology means an increasing number of centralized processing facilities. Instead of family dairies providing milk to the local population, huge milk factories now supply entire regions. Contaminated milk from a large Midwestern dairy caused 250,000 people to become ill from *Salmonella* bacteria. Hen houses don't contain five hundred birds anymore, they contain hundreds of thousands of birds, and an undetected contamination of eggs can cause infections to break out over multistate areas. Technological advances in keeping huge buildings cool are also conduits for disease. There were no problems with *Legionella* before massive air-conditioning and plumbing systems gave the bacteria a friendly breeding environment. Powerful vents that sweep over vats of water containing *Legionella* spread microdroplets of bacteria into the air within buildings. Breathe in the droplets and *Legionella* comes in as well. Even such innocuous advancements as vegetable misters in supermarkets can be a way of spreading *Legionella*.

With new infectious agents being discovered on a yearly basis, the search for new treatments must remain a high priority. Fortunately, humans aren't the only ones who need to protect themselves against invasion by bacteria. Evolution has led to many organisms making small substances—antibiotics—that kill invading bacteria. The antibiotic penicillin, considered by many to be the miracle drug of the twentieth century, was discovered in a fungus by Alexander Fleming in 1928. Penicillin works by keeping bacteria from building bigger walls around their cells while the cells continue to grow. Without an expanding cell wall, the bacteria literally blow themselves up. Penicillin has no harmful effect on human cells because our cells don't build such walls. The manufacture and distribution of penicillin during World War II was followed by streptomycin in 1944, which was highly effective against tuberculosis. Streptomycin and many other antibiotics block bacterial machines called ribosomes from making proteins. Human cells also have ribosomes, but they are different enough from those of bacteria to be unaffected by this class of antibiotic.

Of the thousands of compounds so far identified that kill bacteria, only fifty don't also harm human cells. Still, with such an arsenal of antibiotics, U.S. Surgeon General William Stewart in 1969 gave congressional testimony that we could soon "close the book on infectious disease." Too bad no one thought to inform the bacteria. Anyone who doubts Charles Darwin's theory of survival of the fittest has only to look at the simple bacterium. Antibiotics work because they interfere with the functioning of important bacterial proteins. To interfere with the workings of a protein, the antibiotic must attach to the protein, like the fitting together of two jigsaw puzzle pieces. If the side of the bacterial protein that connects with the antibiotic becomes altered, the pieces no longer fit together. When this happens, the bacterium is able to ignore the presence of the antibiotic and grow without constraint.

Any time antibiotics are used, there is a risk that one of the millions of bacteria being treated is a mutant and makes proteins that don't attach to the antibiotic.[1] While this mutant bacterium may not seem as dangerous as the various *X-Files* mutants, its potential for

[1] Exactly how mutations affect DNA and proteins will be covered in Chapter 3.

harm is much greater. This bacterium will survive the antibiotic treatment that kills all the nonmutant bacteria. A few days later, the mutant bacterium will have multiplied into billions of bacteria that also ignore the antibiotic. As the mutant bacteria spread from one person to another, deadly infections can result unless other antibiotics exist to kill the bacteria. Ironically, hospitals have become breeding grounds for mutant bacteria due to the presence of so many patients with weakened immune systems and by the widespread use of antibiotics.

There is a risk that antibiotic-resistant bacteria can be generated every time antibiotics are prescribed. Every year in the United States, more than 4 million antibiotic prescriptions are given to people who have common colds or the flu, infections caused by viruses. Antibiotics have absolutely no effect on viruses. In a recent Canadian study, 40 percent of physicians gave antibiotics to patients who demanded them knowing full well that they would have no effect whatsoever. Due to such indiscriminate use of antibiotics, the best weapons against bacterial infections may soon be blunted. For this reason, pharmaceutical companies have stepped up the hunt for new antibacterial compounds by searching for more organisms that have found different ways of killing bacteria.

And where will they find such organisms? In places where creatures are still hiding—deep within tropical rain forests or frozen in amber or in rocks underground.

With so much life waiting to be discovered, whether miles beneath the ground, in the heart of unexplored rain forests, hidden inside other organisms, or in the river down the road, it's no wonder that the perils of what lies hidden is a major theme in science fiction. In *The X-Files*, strange, alien worms lurk below the surface of Alaskan ice and feed off the anger of the hosts they inhabit; extinct mites emerge from old-growth trees ready to try a new delicacy—dried-up loggers; fungal spores that inhabit volcanic rocks find human bodies much more to their liking; bugs in the rain forests have a big surprise in store for those who are looking for the next miracle health product; and let's not forget El Chupacabra, the Mexican goatsucker. The fictional organisms that populate these episodes are alive only within the minds of the scriptwriters, yet the type of havoc that they wreak is not confined to the comfort of your television screen. While Mulder believes aliens lurk behind most unusual incidents, it is the plausible, scientific explanations given by Scully—based on the large number of real emerging pathogens—that make these creatures appear very, very real.

Worms on Ice

INT. MAIN BUILDING [ARCTIC COMPOUND]—NIGHT

> SCULLY
>
> A parasite shouldn't want to kill its host.

> HODGE
>
> No. No. This won't kill you, unless you try to extract it. Then it releases a poison, the black fluid that killed the pilot.

> MULDER
>
> You're saying it's possible this worm makes you want to kill others... That could explain what happened to the first team.

> DASILVA
>
> Or what could happen to us.

—"Ice"

In the *X-Files* episode "Ice," a cold but dedicated group of scientists and their faithful dog are innocently conducting global climate research in northern Alaska. Frozen in time, deep below the icy surface, is a record of the environmental conditions on Earth dating back as far as a few hundred thousand years. As the scientists study their latest batch of ice core samples, it's quite obvious (obvious, that is, to anyone with six years of postgraduate geophysical training) that 200,000 years ago, Java Man was enjoying his morning brew as dawn rose on another balmy day. Unfortunately, the scientists are consumed with generating ancient weather reports and forget to see the movie *The Thing* (either version). They don't realize that ice core samples should never be taken in the vicinity of buried meteors. Soon the only survivor in the remote outpost is their aggressive bor-

der collie. The only clue to their untimely demise: the cryptic message "We are not who we are."

FBI agents Fox Mulder and Dana Scully along with a crack team
of doctors and scientists race to the remote Alaskan base to investigate. Instead of finding the morphing alien monster from *The Thing*,
they discover that the demented dog has bizarre cells in its blood,
nodules reminiscent of bubonic plague, and a worm crawling around
below the skin at the base of its skull. Scully speculates that the unusual cells are the free-living juvenile stage, or larvae, that will eventually develop into the adult worm creature. The worm, in true
X-Files fashion, is unlike any previously identified worm but has features similar to a tapeworm—a scolex (weird head), suckers, and
hooks. The dog blood also contains a high level of ammonium hy-
droxide, matching the high level of ammonia in the ice core sample.
Mulder naturally makes the connection between the proximity of
the buried meteor and the original location of the ice core sample.
Undoubtedly shaken by the discovery of so many dead geophysicists,
Mulder is a little off (about 250 million years) on his speculation
about the age of the ice core sample. Nonetheless, he suggests that
the worm survived being rudely separated from the rest of its planet,
a long meteor cruise, and an extensive sojourn under the Alaskan
ice, because it originated on a planet with a frozen, ammonium-
soaked atmosphere.[2]

Why an alien worm would want to take up residence under the
skin of dogs and people becomes a matter of intense speculation between Scully and her fellow scientists. They determine that the wrig-
gly parasite attaches itself to the hypothalamus, a region at the base
of the brain. Removing the worm has a nasty side effect: death. However, having an alien parasitic worm attached to the hypothalamus
has its own severe consequences. While the hypothalamus comprises
only 1 percent of the human brain, it controls most of the basic and
primal behaviors. Without a functioning hypothalamus, eating,
drinking, sleeping, and sex would all be simply memories.

Before the worm can approach the tasty hypothalamus, it needs
to swim through the blood into the brain. The vast majority of the
brain is protected from foreign substances by a roadblock called the

---

[2]Speculations on life in such an atmosphere will be presented in Chapter 2.

blood-brain barrier. Without this barrier, few brain cells would remain for life's golden years. Brain cells come in limited numbers and are not replaced if carelessly destroyed by minute traces of toxic substances in the blood. The blood-brain barrier filters out most of the impurities in the blood before the blood enters the brain. Fortunately for the worm, the hypothalamus is one area of the brain that can be accessed by an alien worm. The hypothalamus has several regions that are not protected by the blood-brain barrier, since the cells of the hypothalamus must be able to sense the state of the body by checking out everything that's in the bloodstream. Feelings of dying of thirst on a hot summer day come from the hypothalamus sensing a high concentration of salt in the blood. Real parasitic worms can also directly infect human brains, probably by eating right through the barrier. The alien worm could therefore bypass the blood-brain barrier just in time for breakfast.

Scully is correct when she states that the hypothalamus is a "gland that secretes hormones." About ten major glands produce hormones in higher animals like humans. A hormone is a substance that is produced in one part of an organism and then travels to another part of the organism where it can cause certain cells to perform some function. In humans, hormones are one of the brain's methods of communicating messages to the rest of the body. Cells that are supposed to respond to the hormone have proteins called receptors either on the surface of the cell or inside the cell. The hormone and its receptor are like interlocking pieces of a jigsaw puzzle. If the mobile hormone puzzle piece floating around in the bloodstream comes in contact with a receptor puzzle piece on the surface of a cell and finds a precise fit, a series of events occurs inside the cell that usually results in the cell doing something new like producing a new protein. The job of the hypothalamus is to secrete tiny amounts of hormones that travel to the pituitary, a blueberry-sized gland just below the hypothalamus. The pituitary is a major hormone producer, regulating everything from milk production to adult height.

A good example of hormone action is the hormone insulin. Insulin is a small protein hormone made in the pancreas, a gland located just below the stomach. Indulging in a sugar-laden ice cream sundae with mountains of hot fudge, whipped cream, and a candied cherry leads to high levels of sugar in the bloodstream. The pancreas

senses the sugar levels and causes insulin to be produced and released into the bloodstream. Once mobile, insulin attaches to any cell
that contains the insulin receptor on its surface. When insulin and its
receptor meet and fit together on the surface of a cell, doorways into
the cell swing open, allowing all that delicious sugar floating around
in the blood to enter. Once inside the cell, the sugar is converted into
carbon dioxide, water, and the fuel of the cell, ATP.

People with diabetes mellitus are unable to make insulin so cells
are never instructed to open up and let the glucose in. Sugar remains
in the bloodstream, causing water to move out of cells to dilute out
the sugar. High levels of water in the bloodstream cause the kidneys
to increase urine output to remove the excess water along with all
the sugar that should have been used for energy. If a cell doesn't get
enough sugar, the body switches to using fats and protein for energy,
which damages critical tissues and organs. Before 1920, diabetes
meant a substantially shortened life span. Now diabetics can perform
the function of their pancreas, monitoring levels of sugar in the
bloodstream and taking insulin when necessary.

Scully and the other scientists correctly speculate that the unusual aggressive behavior of the ice station's men and dog could be
due to the worm sucking on the hypothalamus. That's because the
hypothalamus does more than just sense body conditions; it's also involved in controlling aggression. Electric shocks in the regions of the
hypothalamus that control aggression can provoke aggressive responses in people, much like those experienced by the outpost scientists before they apparently killed each other. Cancer patients
unfortunate enough to have tumors on their hypothalamus can also
become very aggressive.

If the Alaskan worm has a preference for a particular region of
the hypothalamus, it could invoke a particular type of aggression.
Aggression is generally thought of as either offensive or defensive.
Offensive aggression is fighting for social status. Defensive aggression
is protecting children. The aggressive behavior of the dog and the scientist who become infected are more offensive in nature, in line with
the worm attaching to the hypothalamus in the upper middle portion, close to the front of the brain. Harboring an alien worm with an
appetite for brain could certainly cause the afflicted individuals to
feel that they "are not who they are."

Scully and the other scientists speculate that the worms suck on the hypothalamus to make the host aggressive so that the chemical acetylcholine can flood the brain. Acetylcholine would then be eaten by the worms, which probably have worked up quite an appetite after fasting for a few hundred million years. Acetylcholine is a small chemical called a neurotransmitter, which helps nerve cells carry electric signals. When skeletal muscles are stimulated to move an arm or a leg, it's a matter of nerve cells communicating to the muscle cells. The chemical messenger in this tête-à-tête is acetylcholine. For each muscle cell that is stimulated by a nerve cell, about 10 million molecules of acetylcholine are required.

Just as presented in the episode, acetylcholine is also found in the hypothalamus of aggressive people. It is therefore an accurate theory that the worm, by attaching to the hypothalamus, is making people aggressive so that acetylcholine can be produced for food. While a diet of acetylcholine is not quite the equivalent of a hearty meal of meat and potatoes, the food that some organisms survive on is surprising. Consider the simple bacterium. Normally, antibiotics kill bacteria. Some bacteria, however, find antibiotics to be quite tasty. Two hospitalized patients whose bacterial infections were treated with the antibiotic vancomycin made miraculous recoveries as soon as their antibiotic medication was stopped. Instead of killing the bacteria, the doctors were feeding them. Given this unusual example, it is perhaps less surprising that alien worms might find acetylcholine appetizing.

A brain-sucking worm that alters the behavior of its host—fact or fiction? Actually, parasites that change the behavior of their hosts to better suit their own agenda is a common theme in nature. Many parasites have complex life cycles. They spend their formative years (or days) in one host species, known as the intermediate host. To mature into an adult, they need a second host species, called the definitive host. Since the parasite must move from one host into the next, it must persuade the intermediate host to become a tasty treat for the definitive host. Since it isn't normal animal behavior to want to become somebody else's dinner, the parasite needs to change the behavior of the intermediate host by modifying its brain or central nervous system such that the hapless host doesn't mind jumping into the soup pot of the definitive host.

The choice of a tapeworm as the model for the alien worm in "Ice" makes sense scientifically. Tapeworms are masters at changing their hosts' behavior. For example, intermediate hosts for the tapeworm *Taenia multiceps* are sheep and cows; dogs and wolves are its definitive hosts. The sheep first eats the worm eggs, and the hatched larvae travel from the intestine to the brain through the bloodstream. The worm larvae then snuggle up with the sheep's brain, where they live for about seven to eight months. Sheep behave very strangely when they have larvae on the brain. They become listless and move in tight circles away from the herd. Once separated and incapacitated, the sheep are easy prey for wolves and wild dogs—the desired destination of the parasite. Since humans may also be intermediate hosts for this tapeworm, a trip to your doctor is probably in order if you are tired and have an urgent desire to walk in circles by yourself.

Another case of a parasite taking over the central nervous system of a host is that of the lancet fluke *Dicrocoelium dendriticum*. This worm has a greater dilemma than that of the canine worm just described. Its intermediate host is an ant and its definitive host is a sheep. Sheep don't normally find ants very appetizing so the worm has its work cut out trying to make sheep eat its temporary ant home. Together with a group of its buddies, an immature worm enters an ant and wriggles its way into a part of the nervous system that controls mouth parts and locomotion. The worm literally takes control of the ant's body, causing the ant to climb to the top of a blade of grass. The worm then makes the ant bite onto the tip of the grass and hang on until a sheep comes by and eats it.

If a parasite is affecting the behavior of its host, it would seem like a safe bet that it's interfering with the host's brain or central nervous system. Not true, though, for thorny-headed worms that belong to the phylum *Acanthocephala*. The intermediate host is a little aquatic crustacean called an amphipod. Amphipods normally avoid the surface of a lake as they have an understandable aversion to being eaten by predators. This all changes when they become hosts for the worms. Instead of burrowing for safety under the sand when disturbed, the confused crustaceans swim to the top of the pond. Here, they are easy prey for the definitive hosts—ducks, beavers, and muskrats. How this behavior modification occurs is a real X-File since it doesn't seem to involve either the brain or the central nervous system.

Isolated from the outside world, Scully and her fellow scientists search for a way to kill the worms without killing the human host. This job becomes even more critical when one of them, possibly Mulder, becomes infected by one of the worms. Fortunately Scully makes the dramatic and quite accidental discovery that two alien larvae will kill each other if put together in the same drop of blood. She brilliantly extrapolates that worms of a feather might not want to flock together, and two adult worms might just do each other in if introduced into a single host. This antisocial behavior of the worms puzzles one of the other scientists, for how can there be procreation in the absence of at least a little tolerance between members of the same species? The scientist is reminded that worms (at least some earthly varieties) are hermaphrodites, and can reproduce themselves in the absence of the opposite sex.

Hermaphrodite is a label given to creatures that can produce both sperm and eggs. A large number of organisms are natural hermaphrodites, including many parasites such as flukes and tapeworms, as well as some snails and fish. Animals in nine of the sixteen phyla have the anatomy and the ability to be either male or female. Some hermaphrodites live lives that even Mulder would find strange. Imagine the following opening to an *X-Files* episode: The sun rises on sunny California. Two truly gorgeous multisexed black creatures decorated with yellow and blue stripes meet each other while meandering near the beach. Sizing each other up, one creature makes the decision to eat the other. The diner grabs the dinner, and tries to suck it into its stomach. As the tension mounts, the act of cannibalism fails—dinner is just too big. So instead, the two creatures decide to have sex. The credits roll.

These "if you can't eat it have sex with it" creatures are sea slugs called *Navanax inermis*. Unless one comes upon the other from behind, they really do try to eat each other first and if that doesn't work, they have sex. Like the alien worms, the first inclination of two *Navanaxes* is to eliminate the other. Seeing as these strange slugs live in the ocean off Southern California, perhaps Mulder shouldn't be too surprised.[3]

---

[3]Well, actually, since I grew up on the sand in Malibu and the cliffs of Pacific Palisades, maybe it should be me who isn't too surprised.

It is interesting to speculate on why the hermaphroditic condition evolved in many simple animals but not in mammals or birds. Some parasites like tapeworms spend their entire life inside other organisms, living a bachelor existence while making their hosts miserable. Coming in contact with a tapeworm of the opposite sex would be problematic if only single worms can infect hosts; evolution would therefore favor the worm that was self-sufficient. Being a hermaphrodite also means that you can explore new frontiers and colonize new habitats all by yourself. However, this isn't true of all hermaphrodites. Many if not most animals that are hermaphrodites can't tango alone and therefore need another member of their species for procreation. Of course, they don't need to be too selective. Being both male and female, any other member of their species will do.

The inability to fertilize their own eggs also helps hermaphrodites avoid inbreeding. One has only to look at the Peacock clan in the *X-Files* episode "Home" to understand the dangers, both mental and physical, associated with being related at several different levels.[4] Sometimes during a single mating, hermaphrodites take turns being the male or the female, which must be an interesting experience. It was once thought that hermaphrodites lived longer than males, which would give hermaphrodites a natural selective advantage for evolutionary purposes. More recent studies, however, indicate that the males were engaging in normal macho behavior, which tended to shorten the lives of males living with other males during the experiment.

Scully's discovery that two worms will kill each other provides the answer to saving the member of the team that is infected. When an additional worm is added to the infected person, the two worms finish each other off. What then to do about the one remaining living worm? This dilemma leads to an unusual reversal of roles for our FBI heroes. Mulder wants to keep the worm alive, arguing that research is needed on its genetic structure. Scully, the scientist, wants it destroyed, feeling that the worm is too dangerous to live. The rights of species to survive often conflict with the needs of humans whose lives or livelihood demand their destruction. The eradication of poisonous snakes in the Northeastern United States, the elimination of

---

[4]As discussed in Chapter 3.

wolves from many parts of the country, and the destruction of the habitat of the spotted owl in the Northwestern United States are all due to conflicts between nature and man. In the end, Scully wins the argument and the worms are a threat no more.

Mighty Mites in Trees

Cutting down trees in Washington National Forest is not a job for the fainthearted. Chain saws missing their targets . . . trees crashing about . . . logs with a mind of their own . . . and if that weren't enough, loggers in the *X-Files* episode "Darkness Falls" deserve an additional measure of hazard pay for the unexpected surprise they receive after giving the final death blows to a massive old-growth tree. Unfortunately, none of them live to collect. Tiny wood mites that have made this tree their home for hundreds of years are not pleased when forced to vacate the premises after the tree tumbles to the forest floor. These little mites are not your typical wood mites. Repulsed by light and glowing a bright iridescent green, the mites soon realize that the perfect menu for those long summer days isn't dried-up tree, but rather desiccated human, conveniently wrapped in family-sized cocoons.

Radio messages to the dried-up loggers go unanswered, causing the FBI and park rangers to investigate. Mulder and Scully are shocked to discover that swarms of mites are responsible for killing the defenseless loggers. With no visible signs of UFOs or meteor impact craters, Mulder doesn't believe that the mites are the vanguard of an alien invasion. Rather, he theorizes that the mites represent a species that was probably extinct except for the inhabitants of the now dead tree. Mulder bases his views on the fact that modern-day mites aren't repulsed by light and don't glow in the dark or desiccate and cocoon hapless humans. Mulder suggests that preserved eggs of the extinct mites lay dormant beneath the ground for an untold number of years until unearthed by the eruption by a nearby volcano. Waking from their long sleep, the eggs hatched into larvae, which then crept into the tree through its roots. The mites proceeded to feast contentedly on tree innards for hundreds of years until their home was rudely toppled by the loggers.

For Mulder's theory to be within the bounds of extreme possibility, reviving other ancient eggs that are dormant and snoozing needs to be possible. The oldest eggs that have been revived are crustacean eggs laid around 1630 in a pond in Newport, Rhode Island. Sediment caused by Europeans settling in the area covered the eggs and kept them from hatching. The eggs would still be buried had not Roger Segelken of Cornell University unearthed them, which caused many of the eggs to finally hatch.

A considerable gap exists, of course, between reviving four-hundred-year-old eggs and eggs that are millions of years old. However, creatures have been brought back to life that are far older.

As a high school student, I spent many hours at the Los Angeles Museum of Natural History trying to piece together the skull of a 7-million-year-old horse (before you get too excited, it's not the horse that comes back to life). I vividly remember all those bones lying in front of me week after week like pieces from a giant three-dimensional jigsaw puzzle. It was exhausting work. Intense concentration and much trial and error led to only a few tiny fragments reuniting with neighboring bone fragments each hour. Before I began working on the horse, I thought that being a paleontologist would be exciting and fulfilling work. But after a few years of immersion in one stack of ancient bone fragments after another, I realized that being a paleontologist trainee left me with a sense of frustration. That horse was never going to trot away when completed. Nor would the ancient bones reveal many clues about the true nature of the extinct horse.

For a few scientists who study prehistoric plants and animals, these frustrations are partially assuaged by finding their tiny subjects encased in amber. Amber deposits are found all over the world, the oldest dating back some 320 million years. Sap, oozing out of wounded trees, trapped and mummified an astonishing variety of insects, crabs, scorpions, leaves, mushrooms, and even lizards. If the sap hardened in an environment where there was limited exposure to oxygen, it turned into the colorful, translucent substance known as amber. Amber provides a window into the last actions of the trapped creatures—a tiny leaf beetle preserved in the act of fighting off the sap that slowly engulfed it; a jumping spider clutching the millipede it never got a chance to eat; little fruit flies reflexively laying eggs; and midges enjoying one last romantic fling.

Browsing through the beautifully illustrated book on amber by

David Grimaldi (*Amber Window to the Past*), you can almost envision the centipedes, caterpillars, and lizards quickly scurrying up the nearest tree if released from their amber prisons. However, as lifelike as these trapped creatures look, down to the tiny scales on the wings of moths and the profuse hairs that cover the larvae of owl flies, they are, of course, very dead. These encased animals have about as much chance of coming back to life as do ancient Egyptian mummies. But what is true of the mummified animals may not be true of the tiny creatures that inhabited the insides of the dead animals. What if these endosymbionts and endopathogens are still alive, waiting only for a crack in the amber that travels through their mummified animal host to be free at last?

If these tiny creatures are still alive, then their DNA must be undamaged. The DNA of an organism is analogous to the hard drive of a computer. The DNA contains all the information required for an organism to make or acquire the substances necessary for life. Imagine how well a computer would operate if its hard drive was shattered into thousands or millions of pieces (trust me, you don't need to conduct your own experiment). Intact DNA, which is normally present as a single piece in bacterial cells, forty-six pieces in human cells, or as many as a few hundred pieces in some plant cells, is required for any organism frozen in time to restart its dormant metabolism and prepare to live again.

Scientists have been very interested in studying the DNA of organisms preserved in amber—and not simply to create living dinosaur amusement parks. By analyzing the DNA from ancestors of modern organisms, insights can be gained into the evolution of that species.

The discovery that DNA isolated from animals in amber isn't completely degraded—in other words, it isn't broken into millions of pieces—was established in 1992. Tiny fragments of DNA were sufficiently intact to be analyzed from a 25-million-year-old termite and bee. This meant that insect DNA can survive for millions of years, but apparently not in an undamaged form. The next report on ancient DNA was published in 1993, on the same day that the movie *Jurassic Park* was released. Newspaper headlines proclaimed that DNA from the time of the dinosaurs had been discovered and hinted that a real Jurassic Park might be just around the corner. The papers didn't dwell on the minor detail that the ancient DNA came from an organism somewhat less exotic than Velociraptor—a weevil that inhabited the early Cretaceous period some 130 million years ago. Given the likely size of the audience for ancient-weevil amusement parks, the story soon died. However, scientific interest remains undiminished. Since then, about one third of the attempts to isolate DNA from animals in amber have been successful.

The survival of even fragmented DNA from creatures trapped in amber is astonishing. Amber, being the sap of trees, is organic material composed mainly of carbon, hydrogen, and oxygen. The high oxygen content implies that the environment inside the amber is oxidizing, which leads to the production of many free radicals that are damaging to DNA.[5] After millions of years of contact with oxygen, any DNA should be long gone. However, water is also required for DNA to fragment, and amber resin acts like a desiccant to suck water from the cells of the organisms that became trapped. The lack of water must afford some protection to the DNA and allow it to endure the millions of years of exposure to the destructive tendencies of oxygen.

Although the DNA of amber-encased weevils still is highly fragmented, the DNA of endoparasites or endosymbionts may be much more intact. It is well known that some bacteria and fungi when presented with harsh environmental conditions (such as having your host become mummified in sap) are able to form spores. Bacterial spores keep their fragile DNA in a watertight container surrounded by a thick, protective protein coat. Spores are resistant to conditions such as boiling, radiation, pressure, and chemicals that would mean instant death to an unprotected cell. Extrapolations from modern experiments suggest that spores could survive for several hundred thousand years if surrounded by organic material that protected their DNA from the sun's ionizing radiation, another producer of free radicals. Spores inside mummified insects should get plenty of protection from radiation due to the organic material of the amber and the exoskeleton of the insect. However, it is a far cry from saying that spores might survive for one hundred thousand years to showing that they can survive for 25 million years.

[5]For more on free radicals and their role in disease and aging, see Chapter 5.

But this is precisely what Raul Cano from California Polytechnic State University showed in 1995. In an amazing paper published in the eminent journal *Science*, Cano described extracting bacterial spores from the insides of an extinct species of stingless bee encased in a piece of amber that was unearthed in the Dominican Republic. The spores, which Cano revived and successfully grew in the lab, were from a strain of bacteria called *Bacillus sphaericus*. This was significant, since the same bacteria live inside modern-day Dominican stingless bees.

Naturally, upon hearing the news of million-year-old bacteria growing in a lab in California, many scientists were as skeptical as Scully would have been. For this news to become truly accepted, the experiment needs to be repeated by other scientists. Unfortunately, no one except Raul Cano has been able to revive ancient bacteria, although many have tried. The skeptics prefer to believe that the little bacilli were simply contaminating modern bacteria that just happen to also live in Dominican bees and just happened to enter Cano's sterile chamber in California. These scientists will have even more reason to be skeptical when they read the most recent paper from the Cano lab. The latest work describes the isolation from amber of an ancient version of a bacterial species called *Staphylococcus*. Staphylococci don't form spores, so any revived cells must have survived for millions of years in the absence of a protective protein shell. If true, then the survival powers of DNA are much greater than previously realized.

If ordinary bacteria can be revived after snoozing for millions of years inside amber, might dormant mite eggs also survive under similar conditions? Mites are arthropods, just like spiders. At first glance, mite eggs resemble considerably smaller versions of the eggs you eat for breakfast. Crack open the shell and there's a glob of yolk inside. Mite eggs, however, come in little packets like peas in a pod, and can survive very harsh environmental conditions. If a chicken lays an egg in a hole in late fall and forgets about it, the result will be a dead, frozen chicken egg. But mites, like other land-living arthropods, have eggs that survive the winter. Arthropods make their own brand of antifreeze somewhat similar to the antifreeze used in radiators to keep car engines from freezing. Also, just like bacterial spores that must shut down their metabolism during a long sleep, mite eggs slow down their metabolic clocks when in nasty environmental conditions. In another parallel with the ancient spores, mite eggs can live in a desiccated environment. They are like cacti, able to suck water in without letting water out.

As sturdy as mite eggs are, surviving for millions of years underground protected by nothing except their waxy cuticle shell is not likely. Mulder uses brevity when reciting his theory on the origin of the mites—not too surprising given the presence of desiccated loggers hanging in trees. If he had expanded his explanation, Mulder might have speculated that the eggs survived by virtue of being encased in amber. Some of the one million different species of mites currently sharing our planet are internal parasites of insects. If an insect host became entombed in amber millions of years ago, the mites living inside the insect would suffer the same fate. Any eggs of the trapped mites would be doubly protected from the oxidizing environment of the sap and the long-term irradiation of the sun by the body of the mother mite and the insect host.

Mulder speculates that the eggs were unearthed by a volcanic eruption. Imagine a massive volcano hurling rocks, trees, and amber into the air. While it is tempting to imagine the heat of the volcano melting the amber prison, thereby releasing the trapped eggs, any temperature hot enough to melt amber would surely destroy the eggs. It's not, however, beyond the realm of extreme possibility that the amber prison was flung high into the sky by the force of a volcano and then plummeted back to Earth, slamming into the ground. The force of the impact could cause amber, mummified insect, and mother mite to shatter, releasing the stored eggs. If the eggs were still alive, they could become stimulated to begin the process of development into an embryo—perhaps they would just need some sunlight and warm temperatures; the ancient Rhode Island crustacean eggs required only a bit of fluorescent light and a few degrees above freezing to hatch from their four-hundred-year sleep. Once awakened, the mite embryos would develop into larvae, which, still groggy after such a long sleep, might climb into the nearest tree and stay hidden until their descendants are rudely disrupted hundreds of years later.

While the light-sensitive mites lie hidden throughout the day, during the night it's the humans who need to hide. As Mulder and Scully huddle in the loggers' cabin trying to will the sputtering generator to keep the single bulb lit, Scully muses on how the mites might glow in the dark. While Scully believes that the mites absorb enzymes taken from the bodies that they cocoon, there are better explanations. Fireflies glow in the dark because they can make two items: an enzyme called luciferase and a substance called luciferin. The firefly enzyme luciferase is able to make light by combining luciferin with a second substance, ATP, the fuel of cells. Humans don't make luciferase, so the mites can't be sucking this enzyme out of humans. However, humans do make and consume about four pounds of ATP every hour. It's possible that the mites make both the enzyme luciferase and the substance luciferin but not enough ATP. So maybe the mites were sucking the ATP out of human cells in order to supplement their own stores of ATP and keep glowing.

Mulder and Scully survive the night in the loggers' cabin but are attacked by the mites the following night as they try to escape from the single-minded swarms. When they are recovering in the hospital from their near fatal desiccations, a doctor tells Mulder that they found a large concentration of luciferin in their lungs, indicating that the mites probably were producing light using the enzyme luciferase. Although Mulder and Scully recover, the mites are not so fortunate. Teams of exterminators spray insecticide throughout their mountain site, wiping out the ancient swarms and returning the mites to their previous extinct state. It's doubtful that anyone, even Mulder, shed any tears over the loss of this particular species from the planet.

Life on the Rocks

INT. ENTRANCE AREA—NIGHT

MULDER
I've been going over Trepkos' work. Fragments mostly, but I found several references...to a subterranean organism.

SCULLY
What are you talking about?

MULDER
An unknown life form... existing inside the volcano.

Scully regards him skeptically, as he continues:

MULDER
I haven't found anything yet describing the organism in specific terms, but—

SCULLY
Mulder, nothing can live in the volcanic interior. Not only
because of the intense heat, but the gases would be toxic to
any organism.

MULDER
Look at this...

—"Firewalker"

You might think that an active volcano would be about the worst place to discover the latest in new and interesting life. Surrounded by blistering heat and toxic gases, even the most hardy organisms would prefer more hospitable niches. Still, volcanoes don't have to contain life to be interesting. They provide a window deep into the heart of our planet—an opportunity to study the composition of the Earth's mantle, probably all that remains of the chaotic early days after planet formation. And then, of course, there is the little problem of the Earth spewing acres of ash, lava, and gases out of volcanoes and onto devastated countryside. The worst natural disaster of the past ten thousand years was the unexpected 1815 volcanic eruption of Tambora in Indonesia that killed over ninety thousand people.

With over 1,500 active volcanoes, trying to predict the next catastrophic eruption falls on the shoulders of volcanologists and seismologists. These scientists use a variety of instruments to monitor the swelling and rumblings of volcanoes prior to eruptions. To perform the more risky jobs, the latest innovation in volcano research involves using million-dollar robots equipped with video cameras, mechanical arms, and no sense of self-preservation. These robots descend into the mouths of active volcanoes, take readings of the local atmosphere, and bring back the cooled remains of bubbling-hot magma—in other words, rocks. One such robot carries back rocks with an extra, added bonus to the volcanologists in the *X-Files'* episode "Firewalker": tiny hidden fungal spores. Spores that are just waiting to enter the warm and nurturing environment of a host, where they can divide and develop into a mature fungus. And human beings fit the host bill just fine.

The spores present a puzzle—how do they survive the inhospitable conditions of an active volcano? Mulder and Scully find handwritten notes from one of the volcanologists, who also has a sideline interest in slaughtering everyone in sight. The notes diagram the possible biochemistry of the unusual fungus, and suggest that the fungus's biology is based on silicon. Scully is not impressed, believing that the diagram is pure science fiction. Mulder, who has spent quite a bit of time thinking about alternative life-forms, reminds Scully that although all known life is based on carbon, silicon is the next best thing to carbon. Mulder, like many students of high school and college chemistry, probably spent hours staring at the periodic table of the elements that decorates all chemistry classrooms. His photographic memory recalls that right below carbon on the chart is silicon, meaning that silicon has properties similar to carbon. Is Scully therefore being carbocentric not to believe in the possibility of silicon-based life?

All life encountered so far relies primarily on just ten of the more than one hundred known elements: carbon, oxygen, nitrogen, hydrogen, potassium, calcium, magnesium, iron, phosphorous, and sulfur. Silicon is curiously absent from this list, even though it's the second-most-abundant Earth element by weight. Silicon makes up almost 28 percent of the total amount of elements on Earth, right behind oxygen's 47 percent. Carbon is 150 times less prevalent than silicon, at just 0.2 percent, yet it's so important to life that most molecules containing carbon are termed organic and those without carbon are called inorganic.

So why is carbon the fundamental element of life and not the much more abundant silicon? Michael Dewar and Eamonn Healy have spent even more time than Mulder pondering this question and have come up with the following support for carbon- over silicon-based life. A large number of the chemical reactions that routinely occur in all organisms release energy and are called exothermic. If a reaction releases energy, then its occurrence is favorable since the products of the chemical reaction are more stable than the original molecules that engaged in the reaction. Think of an exothermic chemical reaction as a car teetering on a precipice of a mountain. It is favorable for the car to descend to the bottom of the mountain—no energy or gas is required. The car going down the mountain gives off heat—the tires become hot in contact with the ground. The "reaction" of the car ending up on the ground is analogous to an exothermic chemical reaction.

If every possible exothermic chemical reaction in your body occurred right now, you would be starring in your own, real X-file as someone who spontaneously combusted. Fortunately, even the most exothermic reactions in your body require the presence of enzymes to help them get under way. Enzymes are proteins that help jumpstart chemical reactions. That car teetering on the precipice needs a push to begin moving down the hill. Similarly, a lit match is needed to help ignite a piece of paper. Once the paper is set on fire, it will burn on its own and heat will be released. Pushing the car or touching a lit match to paper are actions analogous to those that enzymes perform. Enzymes therefore create favorable conditions so that chemical reactions can take place. Without matches, many millions of lifetimes could be spent staring at a piece of paper waiting for it to spontaneously combust. Without the proper enzyme to catalyze a chemical reaction, the reaction would also take a substantial amount of time to occur.

Silicon-based life would have to face the constant problem of spontaneous combustion. Simple carbon-based chemical reactions, such as combining carbon tetrachloride with water to generate carbon dioxide and hydrochloric acid, occur very slowly at room temperature in the absence of an enzyme catalyst. The analogous silicon-based reaction, combining silicon tetrafluoride and water, results in a violent explosion all by itself.

Such differences in the chemical reactivity of silicon and carbon are probably due to the sheer size of the silicon atom. Silicon is approximately one and one half times larger than carbon. The large size of silicon means that electrons have more room to pack around silicon during a chemical reaction than they do around the tiny carbon atom. This allows chemical reactions to occur more readily. The large size of silicon atoms would also pose a second problem for silicon-based life. Silicon organisms would be much larger than carbon-based organisms with the same number of atoms, making them heavier and not as energy efficient.

Evolution may also have bypassed silicon when designing life because of the greater variety of molecules that can form using carbon compared with silicon. Strong chemical bonds form between two atoms when they share electrons. Both carbon and silicon have four electrons available for sharing and therefore both can be tightly bonded to four different atoms. While this property allows great versatility in the number of molecules that can be formed, the smaller carbon atoms tend to form long chain molecules while silicon prefers more compact three-dimensional networks. Long chain carbon compounds play crucial roles in life. They are the fatty acids that make up the membranes that surround all cells, hormones that allow communication between cells, and vitamins that help enzymes function properly. Without the ability to form long chains of carbon—the technical term is catenation—silicon-based life would appear very different from carbon-based life.

Another reason for the popularity of carbon in living organisms is that carbon participates more easily in certain types of important chemical reactions. All the molecules of life need to be assembled from simple building blocks, much like a house is assembled from basic components like wood and glass. Some of these building blocks are carbon oxides (molecules containing carbon bonded to oxygen). These molecules can easily react with other molecules (in the presence of enzymes, of course), allowing carbons to be added or removed from molecular chains. When silicon is bonded to oxygen, it becomes glass or sand, which, as solid crystal substances, cannot participate in any further chemical reactions.

For all of these reasons, carbon-based life wins out over silicon-based life hands down. If silicon-based life developed in the same places as did carbon-based life billions of years ago, competition with carbon-based life would have soon confined silicon organisms to the evolutionary dustbin. But what about those regions of the planet that are too unpleasant for carbon-based life? Such as deep inside volcanoes? Mulder tells Scully that a silicon-based life-form in the deep biosphere is one of the holy grails of modern science. Mulder doesn't add that it would probably not be any kind of recognizable life. Scientists such as Thomas Gold at Cornell University speculate that silicon-based chemical systems not discernible as "living" may be located in regions deep within the planet, below where carbon-based life can survive. But if silicon-based life-forms exist far below the realm of carbon-based life, they would probably get passed over as just some uninteresting rock.

I am in agreement with Scully that it is highly improbable that the fungus in "Firewalker" has silicon-based biology. However, sand is left behind in the lungs of the dead volcanologists following the emergence of the fungus from its human host. How, Mulder reasons, could sand—silicon dioxide—have gotten in the lungs if the organism isn't silicon-based? An answer to Mulder's question that is less biologically earth-shattering does exist. Many organisms, both past and present, absorb silicon from the environment to make little skeletons. One class of these organisms, radiolarians, have populated the planet for about 600 million years. Radiolarians are tiny marine creatures composed of a single very large cell up to a quarter inch in diameter. Their sturdy silicon skeletons with many extruding spikes make wonderful fossils. Diatoms and silicoflagellates are other single-cell marine creatures that coat themselves with silica to form lovely little opaline silica skeletons. Like these marine creatures, the fungal spores in "Firewalker" could have absorbed silicon from their rocky home and secreted sand into the human host. While not quite as exciting as a silicon-based organism, many scientists would love to (carefully) study a fungus with such unusual properties.

While Scully works desperately to culture the fungal spores, she discovers to her relief that the spores may not survive outside a host. If this result holds up, then maybe she and Mulder aren't contaminated by being in the vicinity of the infected volcanologists and will live to solve more X-Files. Although Scully experiments by adding an undoubtedly delicious broth of human tissue, blood, saliva, and sulfur to the spore's new test tube home, they still stubbornly refuse to grow.

Scientifically, this is a very accurate picture of endoparasites, that class of organisms that spend their lives inside other organisms. It's notoriously difficult to find the right mix of ingredients and environmental conditions to make endoparasites thrive outside their hosts. Scully hypothesizes that the spores must be inhaled immediately after the mature fungus rips through the neck of its previous human host. This hypothesis, while comforting to the uninfected, might not be accurate. The spores could still be alive and infectious and just not willing to grow in a test tube. After all, the spores are able to survive inside a rock deep within a volcano until a new home comes along. The only scientific way to make sure that the released spores are harmless is to try and use them to infect another person. Any volunteers?

Then again, maybe Scully is correct and the spores cannot live outside a host. An organism more used to the typical volcanic gases of water vapor, carbon dioxide, and sulfur dioxide might find the oxygen in the air to be toxic. About half of all known life on Earth can live without oxygen. For some of these creatures, most of which are bacteria, even the slightest trace of oxygen is toxic. Other microbes are more nonchalant about oxygen—if it's around they will use it; if not, they live happily without it. Higher organisms like animals have no choice. You, for example, will breathe in oxygen about 8 million times this year. As the old adage goes, "In comes the good air, out goes the bad air." Oxygen comes in, and carbon dioxide goes out. The oxygen inhaled is used for a specific purpose—to make ATP, the fuel molecule of the cell. A rather complicated series of chemical reactions take place in all cells involving the transfer of electrons from one molecule to another, like an old-fashioned fire brigade transferring pails of water from one person to the next. The final molecule that accepts the electrons is oxygen, which then combines with two hydrogen ions to form water. The oxygen you breathe in is therefore converted into water. This process is coupled (in a way not completely understood) to the making of ATP. The electrons that get handed over to oxygen originate from the food you eat, which is how food is used for fuel. The carbon dioxide exhaled is produced as a by-product of many different chemical reactions that take place in cells and is not directly related to the oxygen inhaled.

Oxygen makes up 21 percent of the atmosphere, but this wasn't always true. Oxygen first appeared 3.5 billion years ago as a highly toxic waste product of a newfangled process called photosynthesis. It took about a billion years for oxygen to become plentiful in the air, thanks to photosynthesizing plants like algae. Besides rusting out the rocks on the planet's surface, oxygen wiped out any microbe that couldn't handle its destructive forces. Those that survived produced enzymes that could neutralize the effects of oxygen or were buried deep underground where oxygen from the surface couldn't penetrate. In the fungal spores in "Firewalker" are allergic to oxygen, they would certainly die in the air unless inhaled quickly by a convenient host.

"Firewalker" ends with Mulder and Scully enjoying a month-long vacation in a level 4 decontamination chamber designed to make quite sure that they were not infected with the lethal fungus. This facility, called the Slammer, is real and is located at Fort Detrick in Maryland. With just enough beds for Mulder and Scully, accommodations include epoxy-sealed cement walls, individual bathrooms, a window that overlooks a grassy courtyard, and a special air supply that is exchanged fifteen times hourly. The staff wait on guests in full space suit regalia and must take decontamination showers and step into an ultraviolet light box when entering and leaving. To reserve free accommodations in such luxurious, and no doubt enormously expensive surroundings, one needs only to become infected with Ebola or Lassa fever viruses or deadly fungal spores from volcanic rocks or alien Alaskan worms.

It's a vacation spot that people are, literally, dying to get into.

Fungi, Fungi, Everywhere

INT N.D. SEDAN

SCULLY
Did he tell you what happened?

MULDER
Flash of light, yellow rain, Maria, Maria!

Scully looks at him, unsure of him for a moment.

SCULLY
He didn't kill her, Mulder.

Mulder gives her a questioning look.

SCULLY

I examined the body of Maria Dorantes and believe the cause of death was natural, if not strange. She seems to have succumbed to a massive fungal infection.

MULDER

A fungus?

—"El Mundo Gira"

Migrant workers in the San Joaquin Valley of central California have many worries—exploitive employers, immigration officers, exposure to pesticides . . . and El Chupacabra, the legendary goatsucker. In the *X-Files'* episode *"El Mundo Gira,"* the Buente brothers, Eladio and Soledad, have problems that are more mundane. Both have fallen in love with Maria, a fetching young woman who takes care of the goats. Suddenly, a bright light flashes and hot yellow rain descends from a cloudless sky. A goat and Maria, her face eaten away, are found dead in a nearby field. Eladio, the brother with Maria at the time of the unusual downpour, suspiciously flees the scene.

Mulder brings a reluctant Scully to investigate the slayings, saying that he is intrigued by the yellow rain, since rainstorms with interesting tints are linked to alien encounters. Scully prefers to believe that the deaths are more prosaic; Eladio must have killed Maria since she preferred his brother. Scully doesn't speculate on who killed the goat. A third theory is espoused by an older woman at the camp: Maria and the goat were killed by El Chupacabra, a mythical four-foot-tall creature with red eyes, fangs, and gray skin that revels in sucking the blood out of livestock and pets.

Scully revises her original hypothesis when she sees that the body of Maria is blanketed with a green-gray fungus. And not an alien, mutant, buried-in-a-rock-near-a-meteorite fungus. Just simple, ordinary *Aspergillus*, a brown mold so useful to humans yet so harmful as well. Scully is probably aware that when enjoying a refreshing soft drink, she is drinking citric acid produced by vats of the *Aspergillus* mold. A different species of *Aspergillus* is used to give soy sauce its tangy fermented flavor. Scully is correct when she tells Mulder that *Aspergillus* is everywhere—in compost and dead leaves, on walls and in household dust. Start an air conditioner in the summer and thousands of *Aspergillus* spores blow out of the vents.

Aspergillus is a scourge of crop plants, especially corn and peanuts. The warnings about not eating peanuts with any sign of mold are given for good reason. *Aspergillus* that infects peanut plants can produce a cancer-causing substance called aflatoxin. Aflatoxin was discovered in England in the 1960s during a turkey disease epidemic. Turkeys were dying and no one could come up with a reasonable cause. The so-called "Turkey X disease" turned out to be caused by an *Aspergillus* contamination of the peanut meal fed to the turkeys. There was so much aflatoxin in some batches of ground peanuts that the poor turkeys were dropping like flies.

Scully tells Mulder that *Aspergillus*, while normally harmless to humans, can be occasionally dangerous, especially when infecting people with weakened immune systems. Every day, your body fights an invisible battle with billions of disease agents like bacteria, viruses, fungi, and protozoa—pathogens that aren't interested in peaceful coexistence. Most of the battles are won since evolution has favored people with top-notch defenses. Some of your defenses work against all types of potential intruders. The defensive system known as the skin is your castle wall. Keep it intact and you stand a good chance of keeping the enemy out. If the wall is breached by a cut or splinter then the enemy gleefully marches in. Of course the enemy can enter through eye or nose doorways or through that great portcullis called a mouth. A castle wouldn't leave such obvious entryways unguarded, and neither does the body. Each of these entries has additional defenses. Tear ducts secrete an enzyme called lysozyme that breaks apart invading cells like bacteria and protozoa. The mucous membranes in the nose trap foreign invaders and secrete lysozyme as a backup defense. What you eat goes into the stomach, an environment not friendly to pathogens. Next time you get annoyed at acid indigestion, think about what that acid is doing to the enemies within.

Even the most secure castle needs mobile defenders, which are plentiful thanks to your immune system. Some of these defenders will respond to anything they think is an enemy without caring who the enemy is. These are the white blood cells called macrophages ("big eaters") that devour invaders whole. At the site of a breach in the skin caused, for example, by a cut or a splinter, an inflammation battle rages. Some injured skin cells fight the invading hordes by making the chemical substance histamine. Histamine enlarges local blood vessels, allowing more blood to race to the site bringing along those hungry macrophages. Extra blood also heats up the region, making the site less hospitable for the enemy microbes to colonize.

If these steps are unsuccessful in repulsing the invasion, 7 trillion elite castle warriors are ready and willing to spring into action. These are the white blood cells called lymphocytes. Lymphocytes come in two classes, T-cells and B-cells. Like all white blood cells, T-cells and B-cells are born inside bones. Immature T-cells learn their battle skills in the thymus (hence the appellation "T"), which in humans is located near the heart. Mature T-cells emerge from the thymus covered with receptor proteins on their surface in an almost infinite variety of puzzle piece–like shapes. The receptors on each individual T-cell are identical; different T-cells have different receptors. If an invading microbe or virus has a protein puzzle piece on its surface that happens to fit into one of the receptor puzzle pieces on a T-cell, the T-cell can turn into a killer cell, spelling doom for the invaders.

Some T-cells don't become killers themselves but instead help their B-cell cousins. B-cells were first identified in the bursa of fabricius (hence the term "B"), an anatomical structure found in birds. B-cells also contain a seemingly infinite variety of receptors on their surface. When the receptor on a B-cell makes a perfect fit with the puzzle piece on the surface of an invader, then that B-cell reproduces itself into an army of identical B-cell warriors. The warriors fight with weapons called antibodies, which the B-cells make and spit out into the bloodstream at the rate of two thousand antibody molecules each second. The antibodies are like loose puzzle pieces with two sides that each can fit with the corresponding puzzle piece on the surface of the invading pathogen. If the antibodies floating around in the bloodstream bump into the enemy, they form clumps with the pathogen, providing a delicious meal for the macrophages. Some of the activated B-cells become memory cells. Years later, if the memory cell encounters the original puzzle piece on a pathogen trying once again to invade, they can quickly spring into action. This is how vaccines work. Disabled pathogens are provided to your immune system in advance, stimulating production of memory cells and making your body ready for when a real attack takes place.

Sometimes it seems as if the immune system has switched sides and joined the enemy. Allergic reactions occur when some harmless substance like dust, pollen, or feline saliva on cat hair triggers a massive and unneeded immune response. The surfaces of these common materials can resemble puzzle pieces that the immune system is primed to respond to. The body initiates a substantial and unnecessary counterattack, complete with T-cells, inflammation, increased histamine levels, and dilation of blood vessels. If only the nose is affected, it's called hay fever. If it involves the lungs, it's asthma. The immune system can also fail to distinguish cells that belong to a person's own tissues and those which are enemy cells. Autoimmune disorders are diseases in which the immune system turns traitor and attacks the body's own cells. Almost 2 million people in the United States are living with lupus, a disease in which the immune system attacks the skin, kidneys, lungs, heart, and brain. Muscle cells are the unwitting target in myasthenia gravis, another immune disorder.

Scully tells Mulder that people with weakened immune systems, who are referred to as immunocompromised, are particularly at risk for infection by any potential disease agents and this may be why Maria died from simple *Aspergillus*. If Maria has a depressed immune system, then the castle that is her body has few defenders. If this is true, then when a potentially dangerous organism like *Aspergillus* bypasses the castle wall, it's free to colonize at will. One of today's frightening realities is that more and more people have weakened immune systems, which is contributing to an explosion of diseases caused by bacteria, fungi, and viruses. In the United States alone, more than 120 billion dollars are spent each year on the treatment of infectious diseases. Between 1980 and 1992, deaths from infectious diseases rose 22 percent, not including the AIDS epidemic.

Maria could have become immunocompromised in one of three ways. First, she could have been born that way. A very small percentage of people have inherited a weak immune system and are naturally more susceptible to infections. Second, she could be infected with the human immunodeficiency virus (HIV), which accounts for a substantial number of people with weakened immune systems. HIV is so deadly because it attacks T-cells, the very system created to combat viruses. Third, Maria could have been exposed to, or treated with, a compound that suppresses her immune system. People who are undergoing chemotherapy or organ transplantation are given chemicals or treated with radiation that temporarily suppresses the body's natural immune responses.

If Maria's immune system was compromised, then a fungus that is normally harmless like *Aspergillus* could have invaded her sinuses and caused a lethal form of meningitis or it could have colonized her lungs, also with fatal results. Scully speculates that Maria's immune system may have been damaged by exposure to high levels of the pesticide methyl bromide. With 27,000 tons of methyl bromide dumped on United States soil each year, exposure of farm workers is a major health problem. Use of methyl bromide also has a second side effect—it causes massive depletion of the ozone layer. Since the ozone layer is all that stands between the sun's UV radiation and people riddled with wrinkles and skin cancer, the Environmental Protection Agency has decided to phase out the production and import of methyl bromide (but not its use) starting in 2001.

However, methyl bromide exposure turns out to be a false trail: regardless of the levels of the chemical in Maria's blood, the prodigious amounts of *Aspergillus* in her system would have been deadly even without a compromised immune system.

Maria is not the only victim of normally innocuous fungi. All the victims have one common denominator: they all came in contact with Eladio. When Scully consults an expert on fungi, known as a mycologist, she learns of a man killed by a dermatophyte, a fungus that causes the relatively harmless but annoying malady of athlete's foot. Apparently, something or someone is causing "harmless" fungi to grow at hundreds of times their normal rates, allowing lethal levels to contact people. The scientist shows Scully an unusual enzyme that he has isolated from several of the speedy fungi, an enzyme that just happens to be the same color as the original yellow rain. When he adds a pinch of the enzyme to *Puccinia graminis,* the fungus that causes the widespread crop disease black stem rust, the fungus immediately starts overflowing its container. The scientist's conclusion: the strange, yellow-colored enzyme accelerates the growth of fungi. The mycologist and Scully are horrified by the implications of an enzyme that turns common fungi into dangerous pathogens. Fungi are everywhere, on everything and everyone. As a scientist, Scully knows that life on this planet won't survive if faced with the prospect of runaway fungus growth.

The growth of fungi at the rate depicted by the mycologist in *"El Mundo Gira"* is pure science fiction. For fungal cells or any other cells to divide, they first need to double all the items in the cell. This means that all the proteins, sugars, DNA, lipids, and other cellular molecules must be duplicated before cell division can begin. In this way, when one cell divides into two, the two daughter cells will be virtual copies of the parent cell. Even the simplest cells, those belonging to bacteria, take about twenty minutes to crank out another set of everything before dividing. The fungus yeast, without which bread couldn't rise, takes a bit longer than bacteria to divide. The super-charged fungi in *"El Mundo Gira"* are dividing in less than milliseconds. It's great science fiction but not quite real science.

Mulder believes that the yellow enzyme being unconsciously spread by Eladio is of alien origin. Since no earthly enzyme can speed up cell division to such an extent, he is probably correct. When Chris Carter calls and asks me to come up with real science to explain some aspect of a story line, I always keep in mind the mantra of science fiction: aliens can do almost anything. So maybe alien enzymes can speed up fungal growth by speeding up how fast the fungal cells can divide. While not my idea, it is not outside the realm of remote possibility.

So how might such an enzyme speed up a cell's ability to divide? One way is for the enzyme to make cells want to divide. Proteins called growth factors cause cells to start immediate preparations for doubling in size and then splitting in half. Growth factors are naturally involved in healing wounds. When you cut yourself deep enough to bleed, the cut doesn't stay open for very long. Tiny cell fragments called platelets that float around in the bloodstream soon arrive on the scene. These platelets release a protein called platelet-derived growth factor. Skin cells contain a receptor for the growth factor protein. When the growth factor joins with the receptor on the surface of a cell in the neighborhood of a cut, the cell is stimulated to start dividing to make more cells that will close the gap caused by the wound. If the alien enzyme functions like a growth factor, fungal cells that normally might be nonchalant about dividing would get the message to start cranking out more cells ASAP.

How quickly a cell can divide depends mainly on how fast it can duplicate its DNA. DNA is made up of two chains that wind around each other—the famous double helix. Regions of the DNA known as "origins of replication" are places where the two chains separate from each other, which forms a local bubble in the DNA. The regions where the chains separate provide spaces for special enzymes, called polymerases, to interact with the DNA and start the duplication process. The more origins of replication, the more enzymes can get their hands on the DNA and start doing their job. If the alien enzyme is able to establish millions of additional origins of replication, the time required to duplicate DNA would be measured in seconds, not minutes. Cell division would be more rapid than normal—but still not fast enough. There is a maximum speed that enzymes can work at to complete the tasks of duplicating a cell's ingredients prior to the cell dividing in two. Enzymes are like little molecular machines, and like all machines that have to physically join components together or break them into pieces, there is some maximum speed at which they work. To make any machine work faster, it must be reengineered. Therefore, the alien enzyme would have to modify hundreds, if not thousands, of enzymes in the fungus to speed up cell division to the extent portrayed in the episode.

But, then, aliens can do almost anything.

Mulder theorizes that the yellow enzyme came to Earth on a meteorite. The meteorite could have plummeted through the atmosphere, creating the bright lights seen by the migrant workers. Mulder speculates that if the meteor smashed into a body of water near the migrant camp, scalding water and concentrated enzyme could have rained down upon Eladio, Maria, and the goat. Meteorites like this fictional one have been crashing into the Earth since the original formation of the planet. Hundreds of tons of meteoroids, the name given to meteorites while they are still hurling through space, enter the atmosphere every day. Thankfully, most are small and burn up before reaching the ground. The ones that land, while detrimental to life and limb of those unfortunate to be in the immediate vicinity, are scientifically fascinating as they provide firsthand evidence of the composition of other planets and comets.

Could an alien enzyme have traveled through space on meteoroids as speculated by Mulder? Most enzymes are proteins, and proteins are made up of chains of molecular beads called amino acids. Extraterrestrial amino acids have been found on meteorites, such as the famous Murchison meterorite of Australia, which fell to Earth on September 28, 1969. Residents reported hearing a crackling and hissing sound, while bright orange and red lights filled the sky. The charcoal-black Murchison meteorite then touched down at supersonic speed and splattered all over the tranquil Aussie landscape. The chemical nature of the Murchison meteorite indicates that it may have been the remnant of a spent comet. Just where it acquired its amino acids isn't known.

Meteorites are sorted into three major groups: iron, stony, and ones that are part iron, part stony. Meteorites that are just iron or stone, while interesting to geologists, barely cause a biologist to raise an eyebrow. However, a subset of the stony meteorites are called carbonaceous chondrites because they contain small rounded bodies called chondrules. These meteorites are filled with complex, carbon-containing molecules, including amino acids. All earthly proteins are constructed from twenty different amino acids. The Murchison meteorite contains some of the common earthly amino acids but also other amino acids that are rare or never-before found. The amino acids in the meteorite are in single units, not joined together as in proteins. Nonetheless, if amino acids can survive a trip through space on a meteoroid, it is not a great leap to wonder if proteins could survive as well. The mycologist tells Scully that the enzyme he has isolated is like no other. He may have been referring to the amino acid composition of the protein. If nonearthly amino acids are in the enzyme, its extraterrestrial origin could be nearly guaranteed.

For some reason, the yellow alien enzyme isn't speeding up the growth of fungi on Eladio. Instead, he is a carrier, a Typhoid Mary, spreading the enzyme and its lethal effects wherever he goes. Typhoid Mary is a general term given to people who, although exhibiting few symptoms themselves, can spread a communicable disease. The original Typhoid Mary was Mary Mallon, an Irish immigrant cook. Between 1900 and 1907, over one hundred New Yorkers became infected with *Salmonella typhosa*, the bacteria that cause typhoid, by eating Mary's less-than-sanitary cooking. Several died. Mary was a threat because she refused to stop working as a cook. Every time a new outbreak of typhoid was discovered, there was Mary cooking for the victims. Exasperated city officials finally banished her to a hospital on a Manhattan island, where she languished involuntarily for thirty years. Today, typhoid carriers must register with local health departments and are not allowed to work as food handlers. Unfortunately, the bacteria they are infected with cannot be eliminated with antibiotics.

"El Mundo Gira" is a story of creatures hidden under our very noses that emerge from obscurity to become deadly pathogens. This is not science fiction. Whether or not a new organism will suddenly be found that causes sickness and death depends on a number of factors. Has the organism acquired new abilities to infect people because it has mutated? Have new ways been found for the organism to come in contact with people? The fungi in "El Mundo Gira" emerge from obscurity because they grow so quickly that the host's immune system is unable to stop the infection. The fungi are transmitted because a carrier, Eladio, is able to spread the enzyme that causes the fungi to become so deadly. This is, in many ways, similar to how real emerging pathogens develop.

Eladio and his brother, their faces increasingly covered in fungus that makes them look like humanoid monsters, are not exiled to a hospital like Typhoid Mary. Rather, at the end of the episode, the Buente brothers are seen walking in the direction of their native Mexico. Perhaps their marred appearances will keep them hidden away and therefore unable to inadvertently kill more people. And maybe people who are spared by their self-exile will glimpse them on occasion, and mistake the brothers for El Chupacabra, giving people like Mulder plenty to do in the years ahead.

Bugging the Rain Forests

EXT. RAIN FOREST—NIGHT

As CAMERA PANS a one-man camp, a long extinguished fire, pup tent, we hear a BURST of radio static, then a faint, trembling voice:

DR. TORRENCE (O.S.)

RBP Field Base, come in. Field Base, come in please.

FINDING DR. TORRENCE

slumped weakly against a log in his one-man camp, talking into his radio mike. His flashlight illuminates his sweat-soaked face covered with large, angry BOILS. He is deathly ill.

DR. TORRENCE

This is Dr. Robert Torrence, with the Biodiversity Project, requesting immediate evacuation from sector zee-one-five...

CAMERA PUSHES IN uncomfortably close on Dr. Torrence. The spill from the flashlight casts shadows on his face, making its already grotesque contours appear even more macabre.

DR. TORRENCE

This is a medical emergency. Please respond.

—"F. Emasculata"

Without tropical willow leaves there would be no aspirin. Without the Australian rain forest there would be no corkwood trees and therefore no scopolamine to ease the queasiness of motion sickness. Without the cinchona tree, there would be no quinine to treat malaria. Without tropical plants, there would be no chocolate, coffee, or ice cream. Animals, plants, and fungi are filled with natural products that provide rubber for tires, sweets for palates, and treatments for illnesses. About half of the world's medicinal drugs originated in wild plants. Of the 50 million different organisms that exist on Earth, fewer than 2 million have been identified. Many of the unidentified organisms lie hidden in the dense, wet forests of the tropics.

It is no wonder then that entomologist Dr. Robert Torrence in the *X-Files'* episode *"F. Emasculata"* is slogging through the vast Costa Rican rain forest looking for exotic insects for his employer, Pinck Pharmaceuticals. Rain forests are overflowing with the largest assortment of insects anywhere on the planet. Insects are everywhere— flying though tree canopies so dense that little light reaches the ground, buzzing at eye level under the green mossy layers that blanket trees, and scurrying along the deeply weathered soil of the forest floor. And, let's not forget about the really cool insects crawling around inside dead pigs. Not even large purple, pulsating pig boils spewing out gooey liquid can keep Dr. Torrence from capturing a new insect species—*Faciphaga emasculata*. After sending the promising little bug back to the home company, Dr. Torrence finds himself afflicted with the same ailment as the pig, and experiencing the same fatal symptoms.

Although the Pinck pharmaceutical company is fictitious, all major pharmaceutical companies have teams scouring remote rain forests for interesting new species of plants and animals. Insects like the one captured by the late Dr. Torrence are returned to company labs, where samples of mashed-up bugs are tested to see if they contain any chemicals with possible therapeutic properties. For example, Pinck Pharmaceuticals scientists may have added *F. emasculata* extracts to cells infected with HIV, the virus that causes AIDS, to determine if something in the extract kills infected cells while leaving uninfected cells unharmed. If the scientists are lucky and find that only infected cells die after the treatment, then they need to determine which of the thousands of chemicals inside the bug mash is having the desired effect. The chemical might be anything—a protein, sugar, lipid, or maybe some other cellular molecule. Once the chemical is identified, the scientists search for unwanted side effects. For example, if the miracle HIV-destroying chemical also causes cells to become cancerous, the cure would be worse than the illness.

Next come extensive animal trials where other side effects can emerge, like purple pulsing oozing pustules. If all goes well (and it rarely does), approval can be requested from the United States Food and Drug Administration for the first human tests. Volunteers are given doses that are minuscule at first, then slowly increased. If side effects are still minimal, clinical trials begin to determine if the chemical has any therapeutic effect on people who are sick. If the answer is yes, then a new drug is ready for final approval and marketing. If the answer is no, then a considerable amount of money has been wasted. On average, the process for the discovery and development of a single drug takes about twelve years and 230 million dollars. The number of drugs that fail along the way far outnumber the drugs that are finally brought to market.

The unscrupulous Pinck Pharmaceuticals determines that *F. emasculata* contains an interesting dilating agent that could help prevent heart disease by opening up clogged blood vessels. Hoping to save a few hundred million dollars, the company decides to forgo the time-consuming and expensive process of purifying the active chemical and testing it on animals. After all, if the chemical is harmful to humans, millions of dollars will be saved by skipping the initial steps. Since enthusiastic volunteers willing to be exposed to *F. emasculata* aren't lining up at company headquarters, an anonymous package containing a dead pig leg crawling with insects is sent to a prisoner at a correction facility. Soon, a contagion is spreading out of control at the prison.

When two of the infected prisoners escape, Mulder and Scully are called in to help in their recapture. When they arrive at the prison, they find not only dying prisoners, but a decontamination crew busily burning everything in sight—including evidence needed for their investigation. One of the accidentally afflicted is Pinck Pharmaceuticals scientist Dr. Osbourne, who informs Scully that the insect itself isn't harmful; rather the bug contains an internal guest called a parasitoid, and it is the parasitoid that is causing the disease.

Parasitoids aren't exactly predators or parasites. Predators are animals that eat living organisms but prefer to live outside their prey. Parasites are much smaller than the organisms they attack. They can live inside or outside their host and don't necessarily kill it. Parasitoids are very large parasites that spend at least part of their lives inside their hosts, which they eventually kill by eating its innards. The most common parasitoids are wasps, which lay their eggs inside insect eggs. The wasp eggs hatch into larvae that eat through the larvae of their insect host.

F. emasculata is not the only *X-Files* episode that uses parasites or parasitoids in the plot. The worm in the episode "Ice" would probably be classified as a parasite, since it doesn't directly kill the host. In "Firewalker," the large fungus that derives nourishment from its human hosts and kills them as part of its life cycle could be considered a parasitoid. Parasites play an important role in the plot of "Never Again." People tattooed in the episode, including Scully, suffer from a condition called ergotism, characterized by hallucinations and mood alterations. It turns out that the ink used for the tattoos is made from rye plants that are contaminated with the fungal parasite ergot. Since the hallucinogen LSD is made from ergot, it's understandable that people exposed to ergot see objects that don't necessarily exist. In the Middle Ages, mass poisonings by ergot were common whenever the growing season was cool and wet enough to cause grains like rye, wheat, and barley to become covered with the fungus. Linnda Caporael, a professor of psychology at Rensselaer Polytechnic Institute, believes that the women accused of being witches in Salem, Massachusetts, in 1692 were not possessed by the devil, merely stoned on rye that was infected with ergot. In modern times, ergot has become a major problem by parasitizing sorghum grain, which is fed to pigs and cows. While it is difficult to determine if the animals suffer from hallucinations, they don't like to eat or produce milk while under the influence.

Dr. Osbourne tells Scully that the *F. emasculata* larvae containing the parasitoid make their way into humans through nasal or eye cavities. The parasitoid then produces a toxin that causes the oozing, pulsating boils and lesions. This fictitious disease is remarkably similar to a real disease called myiasis (from the Greek word *myia*, which means "fly"). Myiasis is caused by the invasion of body tissues or cavities by the larvae of insects such as the human botfly (*Dermatobia hominus*), which is found in Central and South America. Victims commonly return from a visit to these areas with what appears to be either nasty boils, infected skin abscesses, or insect bites. Underneath the skin, though, the botfly larvae crawl.[6]

Within weeks of leaving Central or South America, botfly victims find the first purplish lesions on their skin. Then, like *F. emasculata* crawling out of the pig lesions, mature botfly larvae wriggle out of real lesions. Sometimes, instead of emerging, they merely peek out, appearing to some sufferers like "tiny white things with black eyes." The standard treatment if you're suffering from an invasion of the botflies is to apply raw meat or pork fat to the lesion to draw the larvae out. In the United States, the meat used is bacon and the technique is referred to as "bacon therapy." In a paper published by *The Journal of the American Medical Association*, Timothy Brewer and his col-

---

[6]This is remarkably like the behavior of certain alien worms, first discovered in Alaska by a group of geophysicists.

leagues describe how after covering the lesion with bacon for about three hours, the bacon must be removed slowly to keep the larvae from scurrying back into "the safety of their subcutaneous lair."

The main problem in treating myiasis is convincing patients not to manually extract the larvae at home—crushing the larvae while still under the skin can lead to an infection. Sometimes patients suffering from myiasis are misdiagnosed as having a mental disorder. Imagine the conversation—"Doctor, I feel like I have insects crawling around under my skin!" Based on the titles of articles they publish in medical journals, doctors that treat myiasis patients must have a sense of humor. Some favorites are: "An unexpected surprise in a common boil"; "*Dermatobia hominis* in the accident and emergency department: 'I've got you under my skin' "; and "Souvenirs from Belize: the botfly and the screwworm fly." Thankfully, botfly larvae don't contain any toxic parasitoids so human hosts recover rapidly after bug extraction.

While Scully is helping Dr. Osbourne find ways of controlling the contagion, he explains to her that the parasitoid is undetectable in the human bloodstream and therefore is probably unable to reproduce outside its insect host. Since the growth and reproduction of a parasitoid is normally limited to one specific host, it is not surprising that the parasitoid reproduces only while inside *F. emasculata*. The test devised by Dr. Osbourne to determine if a person is infected (prior, of course, to the appearance of purple pustules on the skin) is to let uninfected insects suck on a person's blood for thirty minutes. If any parasitoid larvae are transmitted to the insects then the person must be infected. Scientifically, this seems like an appropriate although unnerving experiment that Scully has to perform on herself.

The emergence of dangerous disease agents from the rain forests is not simply a threat on *The X-Files*. Ecological changes due to dams and deforestation have led to new diseases such as Rift Valley fever and Argentine hemorrhagic fever. Humans for the first time are coming in contact with animals infected with exotic diseases, as did the entomologist in "F. emasculata." Many such animals are infected with viruses that find human hosts much to their taste. The Evandro Chagas Institute in Belém, Brazil, has recently isolated more than 183 new arboviruses, which are transmitted by mosquitoes, fleas, ticks, or lice and can cause deadly infections in humans.

The organization in the United States that is called in when deadly new diseases emerge is the CDC, with its nearly seven thousand people in 170 occupations. The CDC has recently increased its activities in the field of emerging pathogens as a result of recent outbreaks of a new strain of hantavirus in the Southwestern United States in 1993 and the Ebola virus in Africa in 1995. With old and new infectious diseases breaking out around the world, and modern travel breaking down the traditional boundaries of oceans and mountains, the role of the CDC as sentry and doorkeeper is becoming increasingly important.

The pneumonic plague outbreak in India in 1995 showed how the CDC springs into action. While plague cases are rare in the United States, up to forty people each year in Western states are still stricken with the disease. All three types of plague—bubonic, septicemic, and pneumonic—are caused by the *Yersinia pestis* bacterium, which lies hidden in rodents and is transmitted to humans by fleas. Pneumonic plague, a Class 1 internationally quarantinable disease, has the additional distinction of being transmitted through the air. If you incautiously spend time around a person with pneumonic plague, in one to three days you will come down with a high fever and start coughing up blood. Soon, pneumonia will set in, making breathing difficult. If antibiotic treatments are not begun within twenty-four hours of first contact, death is as certain today as it was in the Middle Ages.

The last major outbreak of pneumonic plague in the United States took place in Los Angeles in 1924. A Mexican woman and eighteen of her friends died within a few days of one another from severe pneumonia. Soon afterward, the diagnosis of plague was made. By the time the last rat in the neighborhood had been eliminated, thirty of thirty-two people infected with the bacteria had died. With this outbreak in mind, the CDC paid close attention when in late August 1994, health officials in India reported the first cases of bubonic plague, which sometimes precedes pneumonic plague. By September, over three hundred unconfirmed cases of pneumonic plague and thirty-six deaths were reported in the Indian city of Surat. Two million terrified residents fled the area, bringing the plague to other cities, including Bombay and New Delhi. Countries across the globe started panicking, with some closing their borders to Indian travelers and discontinuing flights to and from India.

The CDC, unwilling to make a pariah out of an entire country, engaged in its own course of action. The game plan involved writing articles and documents for distribution to public health officials and agencies informing them of the dangers and symptoms of pneumonic plague. The media also became involved, and a special plague hotline telephone number was created. All airline passengers coming from India were given information cards that described signs and symptoms of the plague. Physicians were contracted to serve as quarantine officers at all major airports. It was a bad time to feel queasy on a transatlantic flight since flight attendants reported any sick passengers, all of whom were whisked away to hospitals on landing. Passengers seated nearby were also put on antibiotics. In the thirty days that the surveillance was in place, thirteen airline travelers were examined, none of whom had the plague.

While it is common to think that emerging diseases originate in faraway places like the Costa Rican rain forest before they travel to the United States, the opposite can also be true. In 1995, a woman from Seattle was infected while in East Africa with a disease caused by *Hermetia illucens*, the black soldier fly. This was unusual in two respects. First, *H. illucens* was not known to infect living humans, preferring instead rotting corpses, fruits, and vegetables. Second, black soldier flies are found primarily in Florida and California and were introduced inadvertently into Africa.

Mulder and Scully are hard on the trail of the escaped prisoners, who pose a public health hazard every bit as serious as the plague. Since *F. emasculata* is transmitted by simply contacting the pus oozing from the boils, everyone who comes in contact with the prisoner, including his girlfriend, is exposed. Thanks to the efforts of Mulder and Scully, widespread deaths are averted when the prisoner is cornered and killed before he can infect a bus filled with people. One can only imagine the repercussions at Pinck Pharmaceuticals when the inevitable hordes of CDC, FDA, and FBI agents invade their slimy operations. As all real pharmaceutical companies are well aware, government regulations on new drug development are in place for good reasons. For one never knows who will bring back the next innocuous insect from the heart of some exotic place that has a nasty little guest inside looking to step up to new and better hosts.

2

Visitors from the Void

June 8, 2008, 9:46 A.M.; Fort Detrick, Maryland.

Four scientists garbed in bulky space suits sit clustered around an electron microscope in a sterile, BL4 top containment room. Three of the scientists watch intensely as the project director carefully places a drop of amber liquid from a test tube onto a grid. Inside the test tube are microscopic cells removed days earlier from the surface of a nondescript reddish rock—a rock like many that lie scattered along desert roadways. But the desert this rock came from lies 50 million miles from the nearest road.

The scientists are visibly nervous. They had accepted the current dogma that samples retrieved from the surface of Mars contain no living matter. The Martian atmosphere, only 1 percent that of sister planet Earth, doesn't protect the rocky surface from the intense ultraviolet radiation of the distant sun. Water, in its liquid form so necessary for life, exists at the frigid Martian surface only as ice, and even the ice is restricted to the polar caps. Life might be found by drilling deep below the Martian surface, to where water is liquid and kilometers of rock afford protection from the harsh surface conditions. This kind of deep drilling necessitates the presence of man, and landing men on Mars is still only a paper fantasy. Life shouldn't be found on surface rocks that can be retrieved by robots, yet here they are in the room, waiting and wondering.

The director finishes preparing the specimen for viewing and inserts the grid into the vacuum chamber of the powerful scanning electron microscope. The eyes of all the scientists focus on the microscope's video monitor. The seconds that pass waiting for the image to come into focus seem like years. And the wait *has* been years. Four years in preparation for the Mars Surveyor 2001 mission. Months spent waiting for the ship to make the giant leap across interplanetary space. More time for the rover with its Athena instrument payload to obtain core samples from within surface rocks, and for the samples to be launched into orbit. And then years waiting for the Mars Sample Return Ship to dock with the orbiting sample container and ferry the precious rocks back to Earth. Finally, after all this time, more waiting while old arguments were revisited about the need for additional precautions to protect Earth from imaginary hordes of deadly alien microbes—a scenario expounded by nervous politicians who grew up on too many *X-Files* episodes. All in preparation for this moment.

As the image on the video monitor comes into focus, gasps of shock and disbelief permeate the small room. Is it some hideous alien creature on the screen, some terrifying microscopic monster? No. The image on the screen is a grapelike cluster of cells, identical to those of *Staphylococcus aureus,* a common terrestrial anaerobic bacterium. As eyes close and shoulders slump, the worst nightmare of the scientists is realized. A seven-year experiment with a slim but tantalizing possibility of answering the age-old question of life on Mars may be invalidated. All because terrestrial bacteria must have hitched a ride into space on the very instruments used to drill into the Martian rocks. But then again, what if a single genesis produced life on Mars and Earth and the bacteria are actually hardy Martian cousins?

This fictional scenario is a real-life worry for scientists currently working on missions to collect rock samples from Mars and bring them safely back to Earth. Contamination of drill bits with Earthly microbes is more than a remote possibility. It is sufficiently difficult to build and launch ships into orbit without the added problems associated with having to work under completely sterile conditions. Faced with the possibility that sterility cannot be assured, how will it be possible to determine what is alien and what is not?

If alien life is found in our solar system, first contact will likely be with simple cells rather then either little green men or the "grays" so ubiquitous in accounts of alien abduction. Since cells are the centerpiece of life on Earth, it is difficult to imagine an alien life-form that is not based on cells. Living on earth, however, does give us a bad case of biological solipsism. Mix together carbon, nitrogen, oxygen, phosphorous, and plenty of water, wait a few billion years, and what could possibly emerge except cellular life?

All life on Earth is composed of cells. English scientist Robert Hooke used a primitive microscope to observe cells in plant material in the mid-seventeenth century, but his observations remained unappreciated for 170 years until the German naturalist Theodor Schwann correctly theorized that all organisms consisted of cells. The next major advance came in 1858, when German physician Rudolf Virchow suggested that all cells come from other cells. This was a remarkable statement in its day, that life springs from life and not non-life. The problem, of course, was getting people to believe him, for who had not seen mold appear on bread where no mold was before? Soon after Virchow's hypothesis was publicized, French microbiologist Louis Pasteur proved to skeptical scientists that life did indeed spring from preexisting life. Modern biology still considers the cell as the basic unit of life.

Cells existed for at least 3.85 billion years before being discovered by Robert Hooke. Rocks of this age from Western Greenland contain tiny bits of chemicals that are thought to be the remnants of early cellular life. Actual fossilized cells are visible in rocks dating back a mere 3.5 billion years from Australia and South Africa. Since the Earth was formed about 4.5 billion years ago, cellular life took 750 million years to develop from the original simmering primordial soup. These ancient cells needed to be especially sturdy, given the turbulent times on early Earth. It is hard to imagine trying to make a home and getting to the important business of evolving given the exploding volcanoes and constant meteor showers. Some of the meteorite impacts were so violent, they vaporized the upper layer of oceans. These turbulent times ended about 3.85 billion years ago, giving our hardy little ancestors a chance to spread throughout the planet. In short, cellular life pretty much formed almost as soon as the planet settled down enough to permit it.

Modern cells come in so many shapes and sizes that it is impossible to predict what an alien cell might look like. Some cells are very large: a chicken egg is a single cell; some human nerve cells can stretch over three feet, from the base of the spinal cord down to the toes. Some cells are extremely small, such as bacteria, and can be seen only with the help of powerful microscopes. But all cells, no matter what the shape or size, are a liquid compartment surrounded by a barrier called a plasma membrane, which is composed of lipids (the same molecules in fats) and proteins. The membrane separates molecules inside the cell from the environment. It must be strong enough to keep out large predators like viruses yet sufficiently permeable to let in food and water and flush out wastes. Formation of a plasma membrane was the key to the origin of life.

All life is a series of chemical reactions, in which molecules must physically come together so that new molecules can be formed. Molecules trapped in a compartment by the plasma membrane are confined to a small space, making it much more likely that they will interact with one another. Imagine how long it would take to build a house if the building materials are scattered randomly all over the city. Now consider building the same house if all the materials lie within a fence that surrounds the property. Although alien membranes might have different types of lipids and proteins, it is hard to imagine life-forms developing in their absence.

Television aliens are rarely cells unless they are some giant space-faring amoebae sucking up the starship *Enterprise*. Some TV aliens look like giant copies of animals that already exist on Earth. Other aliens look like humans with bad makeup—after all, costumes have to fit the actors wearing them. Screen aliens do things that give away the fact that they're not your average Tom, Dick, or Harry, like emerging from spaceships on the Washington, D.C., mall with giant robot companions named "Klaatu." While some of the aliens on *The X-Files* are just as fanciful, such as the morphing, green-blooded bounty hunter from the episode "Herrenvolk," others are much more subtle, and therefore more plausible. There are alien worms buried near a meteorite in the episode "Ice," and strange microbes in flasks labeled "Purity Control" in the episode "The Erlenmeyer Flask"; the aliens in the episodes "Gethsemane" and "Redux" are not necessarily the ones that resemble refugees from Roswell, New Mexico, but rather unclassifiable cells that begin to divide when placed in a nutrient media; and finally, there is the so-called black cancer organism introduced in the episode "Piper Maru," which changes from dust into sluglike worms that invade helpless human hosts.

On *The X-Files*, differentiation between what is alien and what is merely strange falls within the auspices of Scully and the numerous scientific experts with whom she consults. These researchers use the latest techniques to ferret out the truth of the organisms entrusted to them. It's a dangerous business. Scientists who analyze alien life-forms on *The X-Files* have much in common with red-shirted security men from the original *Star Trek* series—they rarely survive the episode. You would think that after six seasons and so many deaths, scientists would run screaming from the room at first sight of either Mulder or Scully. Yet without these brave and dedicated individuals, the truth would remain hidden.

Although real scientists have not yet faced the question of living aliens, speculation has fueled commentaries for years on such subjects as whether alien bacteria are sailing the heavens on comets or dust, and whether life originated elsewhere and we are all aliens to this planet. Like many of my scientist friends, I love to fantasize about what alien life-forms might be like. As a biochemist, I think more about the insides of alien organisms than the outsides. What would the genetic material look like? Will the proteins be similar? Will similar plasma membranes enclose the cells? Helping to add realism to the scientific investigations of alien organisms on *The X-Files* has been wonderful because I can make some of my own personal speculations come to life.

How to Identify an Alien in Three Easy Steps

INT. GEORGETOWN MICROBIOLOGY DEPT.—LUNCH ROOM—LATE NIGHT

DR. CARPENTER
What you're looking at is a sequence of genes from the bacteria you brought in. Normally, we'd see no gaps in the sequence. But with these bacteria, we do.

SCULLY
Why is that?

Dr. Carpenter takes a deep breath, staring at Scully for a long beat.

DR. CARPENTER

I don't know why, but I'll tell you that my first call would, under any other circumstances, have been to the government.

SCULLY

What exactly did you find?

DR. CARPENTER

A fifth and sixth DNA nucleotide. A new base pair.
(beat)
What you are looking at, Agent Scully, exists nowhere in nature. It would have to be, by definition, extraterrestrial.

—"The Erlenmeyer Flask"

The *X-Files* episode "The Erlenmeyer Flask" begins with a high-speed police chase. The fugitive, however, is anything but routine. When cornered by the police, he moves with lightning speed and strikes with a strength well beyond that of a normal man. These attributes by themselves are sufficient to pique the curiosity of Mulder. Add green blood and the ability to breathe under water and even Scully can't help but be intrigued.

When Mulder and Scully visit the laboratory of Dr. Berubi, Harvard Medical School class of '64 and the owner of the fugitive's car, they find him busy doing experiments with microbes and monkeys. Like any practitioner of the arts of molecular biology, Dr. Berubi's lab is decorated with numerous Erlenmeyer flasks. Whether glass or more modern plastic, these conical-shaped containers designed by German chemist Richard Erlenmeyer have been used for over one hundred years to grow microorganisms.

While Mulder and Scully are questioning Dr. Berubi, they become curious about the microbes being grown in Erlenmeyer flasks labeled "Purity Control." Did these microscopic entities contribute to the transformation of the good doctor's colleague from hardworking scientist into authority-evading, water-breathing, green-blooded strongman? Since Dr. Berubi is decidedly uncooperative, Mulder and

Scully leave but return later to discover that the doctor has swan-dived from an upper floor window using laboratory gauze around his neck instead of the more traditional rubber ropes around the ankles. The suspicious death of Dr. Berubi has increased Mulder's curiosity. What is a dead genetics expert like Dr. Berubi growing in those Er-lenmeyer flasks? And what experiments require so many angry monkeys? Mulder hands a flask of microbes to Scully and asks her to investigate.

The scientific investigation of the cells growing in the Erlen-meyer flask plays a significant role in this episode. Chris Carter, while writing the script, wanted the experiments to be as accurate as possible. Thus began several long conversations between Chris and myself to map out how a scientist would analyze an unknown microbe. Although my expertise is viruses rather then bacteria, I was well aware of the standard series of experiments that scientists perform when faced with an unidentified microorganism. I told Chris that the first step is to grow more of the sample using Erlenmeyer flasks and a nu-tritive liquid the color of weak coffee. Chris was apparently delighted by the name of the flask as this was to become the title of the episode. The second step, I told him, is to examine the appearance of the mi-crobes using the proper microscopes. Once you know what the cells look like, the third step is to analyze their DNA so the microbes can be classified with respect to known organisms.

Tempting some bacteria to grow and divide in the laboratory has been an endless nightmare for many microbiologists. The litany of excuses for why so many bacteria resist enticements to reproduce has been repeated by many scientists: they might be dead—but this can't be true for all recalcitrant bacteria—or they might not be partial to the cuisine. In general, the richer the food fed to difficult bacteria, the sicker they become; starving seems like a natural state for many microbes. The problem scientists face when trying to grow many mi-croorganisms is that some microbes need almost nothing of every-thing—just a few special, nearly impossible to discover ingredients, without which the bacteria refuse to cooperate and multiply.

Since portraying the endless trials and errors of trying to find a proper food source for bacteria would have TV viewers yawning as they frantically reach for the remote control, Dr. Berubi has already solved the mystery of growing the organism. Scully therefore proceeds to the second step, determining what the bacteria look like. She takes the flask of microbes to scientist Dr. Anne Carpenter at Georgetown University for enlightenment.

Invariably, investigations of any unknown microbe require microscopes. But which microscopes? There are so many to choose from. More often than not, the wrong type is used by television scientists. It is the misuse of microscopes, which always makes me wince, that made me so pleased when Chris first called for advice. At long last, a show existed that will use the correct microscopes because the person in charge cares about accuracy.

Take the light microscope, for example. Available as children's toys and routinely used by millions of high school students, these microscopes magnify specimens illuminated by light. The earliest light microscopes were simply upside-down telescopes. The Italian astronomer Galileo, when not examining the heavens, turned his telescope over and was amazed to see tiny bugs expand to the size of locusts. The earliest microscopes contained merely a single lens and were not much better than magnifying glasses. In the late-sixteenth century, the compound microscope was invented. By using more than one lens in combination, it provided dramatically improved resolution and a new world became visible overnight. Intricate details of fleas, gnats, hairs, nettle leaves, and razor edges were marveled over by naturalists using microscopes of cardboard and wood covered with vellum or ray skin.

In 1665, Robert Hooke, perhaps the most accomplished experimental scientist of the seventeenth century,[1] published *Micrographia,* in which he described his observations of the structure of a piece of cork and saw honeycomb divisions that looked like "a great many little boxes . . . the first *microscopical* pores I ever saw, and, perhaps, that were ever seen." Hooke named the tiny compartments "cellulae," the Latin name for "small rooms." The world of cells was born. Later in the seventeenth century, the Dutch amateur scientist (but highly skilled lens grinder) Antonie von Leeuwenhoek made the first obser-

---

[1] Hooke's extraordinary list of accomplishments includes the invention of the respirator, the universal joint, the anemometer, and barometer; authorship of the theory of combustion; contributions to physics, astronomy, mathematics, architecture, and, of course, biology.

vations of single-celled organisms like bacteria, opening up an entire new world of creatures never before imagined.

Light microscopes have undergone a number of improvements over the years. The current resolving power of a light microscope is about 200 nanometers (about one hundred thousandth of an inch), roughly four hundred times better than the human eye. The light that is used to illuminate specimens provides the resolution limitations of these microscopes. Light waves are subject to diffraction; the waves bend around obstacles in their way, causing the visible image to become less focused. What this means is that conventional light microscopes are powerful enough to see individual bacteria, but can't resolve defining features on their surfaces. Light microscopes can be used only to view the few giant viruses, despite what numerous television shows would like to suggest.

One of the most exciting recent developments in light microscopes is the confocal microscope. The confocal scope produces a greatly amplified image using a laser beam sent through a tiny pinhole. By scanning a specimen in the directions of height, width, and depth, a three-dimensional image is built up by powerful computers. I first told Chris about confocal scopes for the episode "Herrenvolk," in which Scully uses one to get a three-dimensional image of her smallpox vaccination scar.

Early in the twentieth century, German scientists saw an analogy between streams of electrons and waves of light. The radical idea was that a beam of electrons could be controlled and focused within a vacuum by magnets and work like the glass lenses of a microscope, only with much higher resolution. The first "electron microscope" was built in 1931. Over the following decades, improvements in design turned the electron microscope into one of biology's most important tools to examine the microscopic world.

There are two kinds of electron microscopes and both have been used in *X-Files* episodes. The first is called a transmission electron microscope. This microscope is used to view molecules like DNA or the interior of cells. The second type, a scanning electron microscope, is used only to see the outsides of objects, like the mutant flies in the episode "The Post-Modern Prometheus" and the alien bacteria in "The Erlenmeyer Flask." Scanning electron microscopes have produced some of the most breathtaking three-dimensional portraits of the very small and brought great insight to the ultrastructure of everything from cells to hair.

The main disadvantage of electron microscopy is that the specimen must be dead. It can provide amazing snapshots of life, but it can't be used to see life in action. Recently, refinements have been made to confocal microscopes such that they now can give electron microscope resolution of live specimens. If any living alien microscopic organisms are ever recovered, this would be the microscope of choice, so that the precious live sample need not be wasted.

I told Chris that the proper microscope to use for an up close and personal view of Berubi's microbes would be a scanning electron microscope. Dr. Carpenter and Scully marvel at the image of the strange cells on the scope's video monitor. When talking with Scully, Dr. Carpenter refers to the microbes as bacteria. Bacteria and similar-looking microbes called Archaea are collectively known as prokaryotes. Prokaryotes are considered by most scientists to be the ancestral cells—the oldest, simplest, and most primitive life-forms on Earth. Prokaryotes are merely a single cell, one compartment surrounded by the plasma membrane barrier. Floating around in the liquid or cytoplasm of the compartment are all the molecules required for life—the DNA genome or genetic material, along with proteins, sugars, lipids, minerals, and many others. The absence of smaller compartments inside a prokaryotic cell is their defining characteristic and what differentiates prokaryotes from cells of higher organisms, known as eukaryotes. When Dr. Carpenter describes the strange cells as bacteria, she must believe that they do not sequester their DNA genomes in a separate cubicle called a nucleus as do eukaryotic cells. Failure to separate the DNA from the rest of the cell is the hallmark of a prokaryote.

In my conversation with Chris, we discussed what Dr. Berubi's bacteria should look like. Scientists that specialize in the study of bacteria find their tiny research subjects to be fascinating and picturesque. I, on the other hand, can be more objective since I prefer plants and viruses to bacteria. While bacteria are fascinating, their appearance is prosaic at best. Since the bacteria in the episode were supposed to be alien, I suggested to Chris that he use a picture of something much more interesting, like plant pollen grains (my working with plants has nothing to do with finding them to be fascinating and picturesque). Pollen grains make wonderful alien bacteria. They are symmetrical like bacteria yet have intricate, complex surfaces with multiple pits and protrusions. I was delighted with the electron microscope picture of pollen that Chris used in the episode. The grains looked eerily alien as the camera focused in on their ornate pits. After the episode aired, Chris and I chuckled over his reading on a fledgling *X-Files* web site that several observant fans noticed that the alien "bacteria" looked suspiciously like pollen.

Since the "bacteria" that Scully and Dr. Carpenter see on the video monitor of the scanning electron microscope do not resemble normal bacteria, Dr. Carpenter tells Scully that she wants to do further studies. My suggestion to Chris was that a scientist would next look inside a cell, since scanning electron microscopy only gives a view of a cell's surface. To begin visually dissecting a cell, Dr. Carpenter could perform a technique called freeze fracturing. After placing the bacteria in a preservative solution, she would flash freeze them in liquid nitrogen. By striking a frozen block containing the cells with a knife, cracks are created that pass through embedded cells. After Dr. Carpenter sprays the cracked surfaces with platinum, the metal cast is removed and an impression of cells in the metal is viewed with a scanning electron microscope. Dr. Carpenter would see an exquisitely detailed, three-dimensional view of the beautifully ordered plasma membrane that separates a bacterium's single cellular compartment from the outside environment. If the strange microbes were like all cells on Earth, she would see that the plasma membrane is composed of two rows of lipids that completely cover the cell. The perfect symmetry of the lipid rows would be broken by scattered proteins that transverse the membrane.

Besides freeze fracturing Dr. Berubi's bacteria to examine their membranes, Dr. Carpenter would have sliced the cells into many thin wafers using a diamond knife. After staining these ultrathin cell sections with lead, she would use a transmission electron microscope to see exquisite details of the interior of the cell. Scully phones Mulder to report that she and Dr. Carpenter have found something astonishing inside the bacteria—tiny cells that resemble chloroplasts. Scully tells Mulder that bacteria such as these haven't existed for countless years. What Scully is describing are bacteria that may have lived more than 1.5 billion years ago—bacteria in the process of evolving into eukaryotic cells.

Eukaryotes such as protists (amoebas), animals, and plants have within their cells a number of small cubicles in addition to the nucleus that houses the DNA. These cubicles are called organelles and each is surrounded by a membrane. Some of the little cubicles resemble intact prokaryotic cells, complete with their own little DNA genomes. This observation caused biologist Lynn Margulis to propose that billions of years ago, a large, hungry prokaryote gobbled up one of its harmless little prokaryotic neighbors and forgot to digest it. The tiny prokaryote (no doubt breathing a sigh of relief) continued to live inside the larger cell. Eventually, the two cells came to agreeable terms and developed a symbiotic relationship, with both cells contributing to the well-being of the new entity. The ingested prokaryote lost the ability to ever live a separate life and became an integral part of the larger cell. Together, the new entity developed into the earliest eukaryotic cell. Mitochondria, found in all eukaryotic cells, and chloroplasts, which reside exclusively in plant cells, are descendants of such poorly digested prokaryotic snacks. What Scully believes she has discovered is that Dr. Berubi's bacteria are in an early stage of symbiosis with the progenitors of chloroplasts. She is correct when saying that such bacteria have not existed on Earth for over a billion years.

Dr. Carpenter conducts a third experiment while Scully sleeps in a nearby room. She analyzes some of the organism's DNA. One of the inescapable features of cells on our planet is that DNA is the genetic material, also called the genome, of the cell. If Dr. Berubi's strange-looking bacteria have a genome consisting of standard DNA, then the DNA would be composed of the same four constituents, called nucleotides, that are found in the DNA of all earthly organisms. Just as a protein is a chain of amino acids linked one after another, DNA is a chain of nucleotides. The smallest chains of DNA are the genomes of viruses, some of which are only a few thousand nucleotides long. A DNA chain in humans can be hundreds of millions of nucleotides long.

DNA rarely exists as a single chain. Rather, nearly all DNA is composed of two chains of nucleotides that coil around each other to form the famous "double helix." The two chains of the double helix are held together because nucleotides in one chain pair up and form weak chemical bonds, called hydrogen bonds, with nucleotides in the sister chain. The four nucleotides in DNA, abbreviated A, G, C, and T, always choose the same partners to pair with. The nucleotide A in one chain always pairs with T in the sister chain. Likewise, C always pairs with G. So if a short region of DNA in one chain is AATC, the sister chain would have the sequence TTAG. This means that if you know the order of nucleotides in one chain of a DNA molecule, you automatically know the nucleotide order in its sister chain.

Since DNA has only four different constituents, scientists in the first half of the twentieth century thought that it couldn't possibly contain enough information to be a cell's genetic material. After all, the genetic material, the genome of the organism, must contain the entire program for producing a unique living creature. The structure of DNA also seemed too simple—merely a uniform double helix regardless of whether the DNA came from bacteria or humans. It seemed much more reasonable that proteins with their twenty different amino acid constituents and their infinite variety of sizes and shapes were the cell's genetic program. Yet a computer program at its most basic level is a language of only two constituents, zeros and ones. The complexity behind programs as intricate as Windows 98 is achieved by the precise order of billions of these two numbers. The complexity of DNA was likewise found to be a function of the order of nucleotides and not the simple number of different constituents

Dr. Carpenter explains to Scully that she has sequenced (determined the order of nucleotides in) one of the organism's genes. If the DNA in a cell is analogous to a computer's hard drive, then genes can be thought of as programs on the hard drive. Genes are the basic unit of heredity. They are the genetic programs that determine everything from the color of your hair to how shy you are around other people. Genes are regions of DNA that get copied into DNA's molecular cousin, RNA. RNA is similar to DNA in that it is also a chain of linked nucleotides, but the nucleotides used to construct RNA are slightly different than the nucleotides that make up DNA. Most RNA molecules are the instruction sheets given to cellular machines called ribosomes, which in turn use the information to construct proteins. The order of nucleotides in thousands of different DNA genes is therefore what determines the order of nucleotides in the corresponding RNA instruction sheets, which get translated by the thousands of ribosomes in a cell into proteins with a particular order of amino acids.

Although proteins are not the genetic material of the cell, they are what directly determines your appearance and basic behavioral traits.

The particular gene that Dr. Carpenter would have sequenced is the one called ribosomal DNA. This gene specifies the production of an RNA molecule called ribosomal RNA. Ribosomal RNA, unlike most RNAs, is not an instruction sheet for making proteins. Rather, ribosomal RNA is one of the numerous cellular ingredients that together make up a ribosome. Since all cells need to make proteins, all cells contain ribosomes. When a new creature is discovered, whether it be a simple bacterium or complex animal, scientists examine the order of nucleotides in its ribosomal DNA gene. Once determined, the sequence is compared with ribosomal DNA sequences of other known organisms. Organisms that are more closely related have genes that share more similar orders of nucleotides. By comparing the ribosomal DNA sequences for all known organisms, a "tree of life" can be drawn, with the earliest organisms at the base, and the most recently emerging organisms forming the outermost branches. If species are near each other on the tree, this means that their DNA sequences are closely related.

An organism unrelated to life on our planet might have a cellular biochemistry that is beyond our imagination. Such a creature would probably not have DNA as we know it, let alone have ribosomal DNA genes. If Dr. Carpenter determines that her strange organism contains ribosomal DNA genes and that the nucleotide sequences of these genes are similar to those found in other organisms on Earth, she would conclude that it was simply a strange new type of bacterium. But this is *The X-Files,* so that isn't what she finds.

When Chris originally talked with me about this script, he needed Dr. Carpenter to discover something about the bacteria that would instantly suggest an extraterrestrial origin. My suggestion was for Dr. Carpenter to discover something completely unexpected when she is sequencing some of the organism's DNA. She would find that Dr. Berubi's bacteria contain DNA with six nucleotides instead of the usual four.

So how could Dr. Carpenter make this startling finding? DNA sequencing is now a routine laboratory procedure. The results of such an experiment are displayed as four ladders, one for each of the four DNA nucleotides, on a piece of X-ray film called a sequencing autoradiograph. With a few minutes of simple instruction, even a child can look at the finished film and determine the order of nucleotides in a sequenced DNA fragment. Although *The X-Files'* audience was unlikely to have a detailed grasp of modern molecular biology, I suggested to Chris that Dr. Carpenter find gaps in the nucleotide ladders on the X-ray film of the organism's DNA sequence. These gaps could be interpreted as evidence of an additional pair of nucleotides—a fifth and sixth nucleotide not found in nature.

Dr. Carpenter explains to Scully (and the home audience) that DNA is always composed of two different pairs of nucleotides, the A/T pair and the C/G pair. Evidence for a third pair of nucleotides in DNA is so unexpected and so different from DNA in any life-form on Earth that she has to conclude the bacteria are extraterrestrial. Chris asked me what I would do if I came across such a result in my own work. I laughed and said that I would probably call the government (not that I ever envision making such a phone call). Dr. Carpenter tells Scully in the episode that normally, her first call would have been to the government. But government agent Dana Scully is already present to share the startling news of the existence of extraterrestrials.

Since no normal DNA-sequencing autoradiograph would contain gaps, I agonized over how such an autoradiograph could be artificially doctored for the episode. I soon realized that this was taking scientific accuracy to a ridiculous level. I suggested to Chris that he visit a molecular biology lab at the University of British Columbia, near where *The X-Files* is filmed, and ask for a standard "sequencing autoradiograph," which I dutifully spelled out. I suggested that Chris have Dr. Carpenter point to the nucleotide-sequence ladders on the X-ray film and exclaim to Scully that she has found gaps in the sequence. This no doubt disappointed the 0.00001 percent of the population who could interpret the real autoradiograph shown in the episode and clearly see that there are no gaps in the DNA sequence.

Dr. Carpenter and Scully find one more intriguing property of the alien bacteria: they are filled with virus. Since thousands of viruses can be present in an infected cell, it is usually a simple matter to recognize if cells are experiencing a virus invasion by using a transmission electron microscope. Viruses can infect nearly all types of cells—bacterial, fungal, plant or animal—so it is not surprising that alien bacteria also contain virus. Although it is a rare cell that doesn't need to worry about virus infection, a single virus is adapted to infect organisms only from one of life's six taxonomic kingdoms. Viruses that infect bacteria cannot infect members of the Archaea (the other type of prokaryote), protists (such as amoebas), fungi, plants, or animals, and vice versa. However, viruses can infect several different hosts within a kingdom. Influenza virus, which causes most cases of the flu, can infect birds, pigs, and humans. The virus that I study, my much-beloved turnip crinkle virus, can infect turnips, mustard, Chinese cabbage, and a lovely little weed called *Arabidopsis thaliana*. Despite viruses' varied appearances and hosts, they all have a single-minded goal: invade a cell, hijack its ribosomes, and force the cell to produce more virus as quickly as possible.

Viruses are simply a set of genes on the prowl. The most rudimentary viruses are merely a protein bag stuffed with one or more pieces of DNA or RNA. Viruses have no independent metabolism. They cannot make proteins or copy their genome, so they must find a cell that is able, if not willing, to help. Relating back to our previous analogy, if the DNA genome of a cell is like a computer's hard drive, then a virus is a bag containing a program on a floppy disk. If there is no way to open the bag, or no computer to insert the disk into, the program is without meaning. Viruses are not cells. The average size of a virus is only about one hundredth the size of a cell. Viruses have no nucleus, no ribosomes, no metabolic activity. They can't absorb nutrients or make proteins. If a virus cannot find a cell to infect, it has no existence.

Since I work on viruses, students often ask me if viruses are alive. I like to answer this question with the question, "How do you define life?"[2] Since this generally leads to vacant stares, I usually issue the reassuring statement that entire books have been written trying to explain the scientific meaning of life.

Most scientists will agree that cells are the basic unit of life, but why? At the very least, something that is alive should have a metabolism, the ability to either absorb or create the materials required for its existence. It should be able to reproduce itself, and evolve into

---

[2]Occasionally this prompts the follow-up, "Why do you always answer a question with a question?" To this, my standard response is, "Do I?"

forms better suited to its environment. The simplest organisms that fulfill these parameters are cells. But as a fan of science fiction, I would argue that the future will likely see a change in this definition. With computer and robot designs becoming increasingly sophisticated, the future could see the emergence of sentient machines with control centers that mimic the human brain. Although no one will argue that an automobile is alive, who will argue that an intelligent and self-aware machine is not as alive as a bacterium?

To invade an animal cell, a virus must trick the cell into letting it through the membrane. Getting most large substances through the plasma membrane barrier and into a cell requires the help of receptors, the puzzle piece–like proteins that sit on the surfaces of cells and serve as gatekeepers, allowing in only those substances that the cell needs. When a specific molecule passes by that can fit together with a receptor, a doorway into the cell is activated. Viruses have cleverly exploited this doorway by having proteins or sugars on their surfaces that mimic the shape of natural molecules in the body which normally interact with receptors. When a virus binds to a receptor, the receptor doesn't realize that it's being duped and opens a way for the virus to invade the cell. Since different cells have different sets of receptors, viruses are limited to infecting only the cells that have a receptor with which they can interact. This is why viruses like the human immunodeficiency virus (HIV) can only infect certain types of cells like immune-system cells and various brain cells—cells that have the virus's receptor, called CD4, on their surfaces.

As the virus sneaks into a cell, the protein bag is opened and the genome of the virus is released. If the virus has a DNA genome, the viral DNA enters the nucleus and commandeers the cell's enzymes to make RNA copies of its genes. The virus subtly alters the cell's ribosomes, causing them to ignore the cell's own RNA and translate only the virus's RNA into proteins. If the viral genome is RNA instead of DNA, the task is even simpler. The viral RNA is directly used to make proteins. Some of the new proteins are enzymes that replicate the genome of the virus. Other proteins are made that form the virus's coat, the bag that surrounds the viral genome. When all is ready, newly made protein bags assemble around freshly made virus genomes. Most animal and bacterial viruses then burst through the superfluous and fatally injured cell and hunt for more virgin cells to infect.

So what was Dr. Berubi doing with an Erlenmeyer flask filled
with alien bacteria replete with viruses and chloroplasts? Scully and
Mulder speculate that Dr. Berubi was using the bacteria to grow virus
and then using the virus to transfer genes into monkeys and his assistant. Using viruses to transfer genes between organisms, even between
kingdoms, is not science fiction.[3] Some viruses with DNA genomes can
insert their DNA into the genome of the infected host. It is not a difficult matter to splice a foreign piece of DNA into the virus's DNA (for
professionals, at least; don't try it at home). Then, when the virus naturally inserts its genome into the DNA of the host, the new piece of
DNA is inserted as well. Dr. Berubi was using bacteria as many scientists do, as a convenient means of producing additional copies of the
genome of his engineered virus. To successfully amplify a virus genome
in bacteria, the virus first must be disguised as an ordinary piece of bacterial DNA. To accomplish this deception, scientists adds bits of bacterial DNA to the virus DNA. One of these DNA bits contains a bacterial
origin of replication. In this way, the bacteria's enzymes think that they
are replicating the bacteria's DNA when they are really replicating the
virus's DNA. Mulder and Scully's idea that Berubi is using the bacteria
as a living factory to grow virus is thus a distinct possibility.

Another piece of the puzzle comes from Dr. Berubi's file folder.
Dr. Berubi was working on the Human Genome Project, which Mulder calls the largest scientific project ever undertaken in the history
of science. Mulder's description of the Human Genome Project is accurate. It is an immense, multibillion-dollar, fifteen-year effort to determine the precise order of the 6 billion nucleotides of the human
genome. It is the twentieth-century equivalent to ordering the one
hundred known elements of the periodic table, except that the ordering involves 100,000 human genes.

Dr. Berubi is not alone in working on the Human Genome Project. This monumental effort involves over 250 laboratories in the
United States along with labs in eighteen other countries. When you
consider that the largest genome sequenced to date is that of the
common baker's yeast, a seven-year effort involving a mere 12 million nucleotides and six thousand genes, the Human Genome Project
seems overwhelming. But by starting with smaller genomes, such as

---

[3]For a more complete explanation, you'll have to wait for Chapter 4.

those of bacteria and yeast, technology has led to more efficient automation of DNA sequencing. Although less than 3 percent of the human genome has been finished so far, scientists believe that they will comfortably make their 2005 goal.

Once the genome is sequenced, scientists and medical doctors will quickly be able to identify genes involved in genetic diseases like some forms of cancer. Individuals that are predisposed to particular diseases will also be identified and treatments will shift to prevention-based approaches. As is revealed in subsequent *X-Files* episodes, the goal of Dr. Berubi and secret forces in the government is to use viruses to insert alien DNA into humans and create alien-human hybrids. Dr. Berubi may have found that only certain individuals are able to survive the introduction of alien DNA. Through his work on the genome project, he could identify which genes give individuals the ability to survive and thus be able to predict who will benefit from the alien DNA and who will die.

While Scully and Dr. Carpenter are working on the strange contents of the Erlenmeyer flask, Dr. Berubi's green-blooded assistant and recipient of the alien DNA is found and carted off to a hospital in an ambulance. While en route the paramedics, fearing their patient is dying, attempt to do a needle decompression procedure. When the needle penetrates the skin, toxic fumes come from the body causing the paramedics to faint.

This scene was written not long after a similar incident became a major news story. In 1994, thirty-one-year-old Gloria Ramirez, now known on the Internet as the "Toxic Lady," was rushed to the emergency room of Riverside General Hospital in California suffering from respiratory and cardiac distress related to her cervical cancer. While drawing blood, a nurse noticed a strange, ammonialike odor and fainted. A doctor took a whiff of the syringe with the patient's blood and also passed out. By the time the incident was over, four additional ER workers had fainted and twenty-eight other people were affected. Gloria Ramirez died in the ER of kidney failure a short time later.

The scene at the hospital must have resembled an *X-Files* episode. The ER was evacuated. Riverside city HAZ MAT (Hazardous Materials) teams dressed in space suit–like apparel took air samples, decontaminated the room, and sealed Gloria's body. If this were a fictitious incident in an *X-Files* episode, Mulder and Scully would have arrived soon afterward. Scully would have conducted the autopsy, finding some scientific basis for the white flecks in Gloria's blood. Mulder would have questioned family members about whether Ms. Ramirez complained of being an alien abductee or whether she recently visited sites of illegal pesticide sprayings. The case would be solved and the X-File closed.

Life, however, rarely imitates fiction. The autopsy of Ms. Ramirez was inconclusive. White particles in her blood could not be explained. There were no traces of ingested pesticides, which have occasionally produced odors in ERs. The hospital was ruled out as a source of the toxic fumes. Mass hysteria or previously undiagnosed medical conditions of the ER personnel were deemed highly unlikely. Dr. Julie Gorchynski, one of the first affected, had been a champion surfer in glowing health. Muscle spasms and oxygen loss following the incident have led to a rare destructive bone disorder. This real X-File remains open and unsolved.

Scully returns to finish helping Dr. Carpenter analyze the alien microbes, only to learn that Dr. Carpenter and her entire family were killed in a car crash. The death of Dr. Carpenter affected me in a rather unusual way. You see, Dr. Carpenter was named after me. Chris had told me that he would name the scientist in the episode Dr. Anne Simon. It was pure bad luck that another Anne Simon was a scientist at Georgetown University, where the fictitious Dr. Carpenter worked. The name therefore needed to be changed to my first name and my husband's last name. Before reading the script, I had visions of Dr. Anne Carpenter as a recurring character providing insightful scientific analysis season after season. My only hope now is that Dr. Carpenter faked her own death and is hiding somewhere, waiting for the opportunity to rise again and help Mulder and Scully in their quest for the truth.

Alien Hitchhikers

INT. U.S. CUSTOMS PRIVATE SEARCH ROOM—NIGHT

                    CUSTOMS OFFICER
Would you mind telling me the kind of diplomatic work you
do, sir? And what material you're transporting in these?

He lifts TWO METAL CYLINDRICAL CANISTERS into view.
ANGLE TO INCLUDE MAN
Reacting to the site of the canisters.

MAN
That is—those are filled with biohazardous material.

CUSTOMS OFFICER
Then where is the paperwork? And why aren't the containers
marked?

MAN
Don't open those. Whatever you do. That material cannot be
exposed.

—"Tunguska"

At the end of long overseas trips, it seems foreordained that my
luggage is searched by United States customs agents. The drill is so
routine. A stony-faced man scatters my personal belongings all over
the counter. Then eyes gleam and a smile tickles his lips when hearing that I am a scientist who works on plant viruses. The inevitable
questions begin. "How many farms teeming with diseased crops have
you visited, Dr. Simon?" "What insidious viruses are you bringing
back into the country, Dr. Simon?" As the minutes tick by and my
chances of making that connecting flight go from dim to nonexistent,
it's easy to forget where customs agents belong on the tree of life.
Still, on these agents' shoulders rests an important responsibility—
protecting the United States from invasion of deadly foreign
pathogens, illegal drugs, and Cuban cigars.

While these are not items that I generally carry in my luggage,
a passenger in the X-Files' episode "Tunguska" is not so innocent.
Despite pleas of diplomatic immunity, agents open his suitcase to
find several suspicious-looking glass tubes containing a
biohazardous substance. While this revelation should have
provoked some care in handling the containers, one slips out of an
agent's clumsy fingers and shatters on the hard floor. Within
seconds, a black, oily powder congeals into tiny, sluglike worms. The

agent gasps in horror as the creatures slither under his pants legs and invade his body.

Unaware of this event, Mulder and Scully receive a tip that they should intercept a rock that is carried by another courier. The black, stony nature of the rock indicates that it is a meteorite. At the Department of Exobiology, Goddard Space Center, Dr. Sacks examines the rock and makes a startling discovery. He tells Mulder and Scully that the rock is indeed a meteorite and one that contains polycyclic aromatic hydrocarbons.

Polycyclic aromatic hydrocarbons, or PAHs, are a set of organic molecules composed of carbons and hydrogens arranged in a ring-shaped pattern. Take a breath in a heavily polluted urban area and your body will have an intimate association with PAHs. The major source of PAHs in the atmosphere is poorly combusted fossil fuels provided by autos and industrial plant emissions. Even minute doses of PAHs can cause cancer. Ever heard the advice against eating too much charcoal-broiled meat? That's because PAHs are released from charcoal, especially mesquite, and absorbed into the grilled food. Vegetables can also take up PAHs from the atmosphere while they are growing, especially large and leafy ones like lettuce and tobacco.

The PAHs that Dr. Sacks finds are in the interior of the mysterious rock, so they did not arise from the meteorite slamming into a barbecue alongside a Los Angeles freeway. Dr. Sacks is excited to find PAHs inside the meteorite because he knows that PAHs are produced by dead, decaying organisms. In other words, PAHs can be a sign that living creatures once resided inside the rock. Dr. Sacks determines that the PAHs in Mulder and Scully's rock are similar to those scientists found in the Antarctic meteorite ALH84001—the Martian meteorite with tantalizing but highly controversial signs of ancient life.

Of the twelve pieces of Mars that have been found on Earth, ALH84001 is by far the oldest—a potato-sized, 4.5-billion-year-old rock that weighs a little over four pounds (1.9 kg). Scientists can say with certainty that these meteorites were once part of the Martian landscape by analyzing gases trapped inside minute bubbles of glass within the rocks.[4] The glass found in ALH84001 is evidence that the

---

[4]While these meteorites were found as far back as 1865, only since the Viking mission have we known enough about the Martian atmosphere to know their origin.

rock was once near a cataclysmic event, a comet or asteroid smashing into Mars with such violence that small parts of ALH84001 melted and formed bubbles of glass. Trapped within the tiny bubbles are whiffs of the atmosphere of Mars, which the Viking mission of 1976 measured to be 95 percent carbon dioxide, 2.7 percent nitrogen, 1.6 percent argon, and a few trace gases.

If you think about it, the presence of Martian rocks on Earth is pretty remarkable. For an object to be ejected from the surface of Mars into space, a comet or asteroid would have had to strike the planet with enough force to throw that object outward at a speed of nearly 4 miles per second—five times the velocity of a rifle. The pieces of Mars so launched would then remain in the solar system (leaving would require a velocity of more than 350 miles per second) for millions of years until captured by the Earth's gravitational field. One of those pieces, ALH84001, wandered homeless through the so-lar system for 16 million years before plummeting into the Antarctic. Thirteen thousand years later it was picked up by someone, only to languish unappreciated for eight more years in a cabinet at the John-son Space Center in Houston.

ALH84001 is not your garden-variety Mars meteorite. Unlike its eleven more youthful siblings, it contains PAHs, the first evidence that organic molecules exist on Mars. This finding came as quite a shock. Over twenty years ago, two Viking space probes conducted experiments that tested for traces of organic compounds on the sur-face of Mars. One test came back positive, the other negative. NASA, not wanting to confuse the public, decided to issue a confident report stating that Mars was a lifeless wasteland devoid of organic materials. But similar positive and negative results were obtained when the two tests were conducted on the surface of Antarctica. NASA chose not to release a further statement that no life existed on Earth.[5] The tests conducted on Earth were not sensitive enough to detect the vast amounts of frozen fungi, algae, bacteria, and diatoms now known to lie deep below the ice surrounding the South Pole. One can only speculate on what the same tests may have missed on Mars.

Besides containing organic PAHs, ALH84001 has many tiny glob-ules of carbonate, a substance that can form when water saturated with

[5]Take note, paranoid conspiracy theorists/*X-Files* fans.

carbon dioxide permeates through rock. Associated with the carbonate globules in the meteorite are chains of magnetite beads—an iron-oxygen compound. Finding carbonate, magnetite, or PAHs by themselves would hardly promote much excitement outside the world of geologists and astrophysicists. But carbonates usually indicate the presence of water, the elixir of life; magnetite beads in precisely the same chain patterns are often found with fossils of Earth bacteria, and the PAHs were of the type produced by decaying microbes. Discovery of all three compounds together within a hundred thousandth of an inch of objects identical to putative terrestrial miniature bacteria, and the tantalizing conclusion reached by respected NASA and Stanford scientists is that life existed and may still exist on Mars.

When NASA held a news conference to announce that ALH84001 contained signs of life, the story was featured on television and newspapers around the world. At my university, microfossils and life on Mars became the major topic of hallway conversations. While some residents of Earth probably wondered what all the fuss was about, others pondered the mind-boggling scientific and religious implications. Life on other planets would mean that we are not alone, that conditions on Earth are not uniquely suited to the development of living organisms.

With all the hype that followed the announcement by NASA, would you be surprised to learn that this was not the first report of fossils in meteorites? In the 1960s, George Claus and Bart Nagy used a hot new instrument called an electron microscope to examine meteorites known as the Orgeuil chondrite and the Ivuna rock. They were shocked to find minuscule shapes that looked like fossils of bacteria. Fearing contamination after impact, Claus and Nagy looked deep into the rocks yet still found the same lifelike structures. Realizing the implication of such findings, Claus and Nagy published a cautious paper in the journal *Nature* suggesting that maybe, just maybe they had found signs of ancient extraterrestrial life.

Claus and Nagy's paper was received with a tidal wave of criticism and indifference. The samples must have been altered after impact; the fossils were artifacts of the examination procedure; or simply, the "microfossils" had to be the result of natural, inorganic processes in the rock, since they were much too small to be real cells. Life on other planets in our solar system was a topic for *Star Trek* fans, not serious scientists. Conditions were too harsh on other planets. Even the best possibility for life outside of Earth, Mars, was too cold and the only surface water was frozen. The moons of Jupiter were too distant from the sun and the presence of liquid water was also thought to be highly improbable.

So why are reports of fossils in meteors being taken seriously today? It has to do with the attention given to the question, Where does life exist on Earth? Actually, the question isn't where life exists but where *doesn't* life exist. There are abundant microbes in the freezing cold of the Antarctic ice. Beneath the burning desert sands are not only bacteria but specialized insects. Twenty-five miles up in the atmosphere float bacteria. At the bottom of the deepest ocean, where no sunlight penetrates, where the temperature hovers right around freezing, and organic nutrients should be too dilute to support anything, exotic zoos of life survive.

But what really led scientists to open their minds to the possibility of fossils in Martian rocks was the abundance of life below our own ground. Microbiologists had once thought that microbes only lived in mud close to the surface. Below the surface layer was solid rock, and what could survive within rock? Yes, there were signs of bacteria deeper below the surface, but the bacteria couldn't be grown in the laboratory, and what couldn't be grown had to be dead or not very interesting.

Then the U.S. Department of Energy decided to check the quality of water below one of its nuclear plants. Deep holes were drilled near the Savannah River in South Carolina. What was uncovered was a world teeming with life, as much life as on the surface, maybe more. Miles beneath the ground, in what has been called the deep, dark biosphere, live bacteria and fungi, basking in the heat of the Earth's core and snacking on rocks and leftover organic material from dead neighbors. The bacteria even make their own organic molecules like methane—not by photosynthesis, which requires sunlight, but from carbon dioxide and hydrogen gas dissolved in the rock. Some scientists even believe that life began under the planet's surface. The abundance of life below our ground has made many scientists reevaluate whether surface searches for extraterrestrial life are giving misleading results.

Dr. Sacks's excitement in finding that the meteorite contains PAHs is therefore understandable—he believes that this meteorite may also contain signs of ancient extraterrestrial life. But PAHs by themselves are only one piece of the puzzle. Dr. Sacks tells Mulder and Scully that he wants to search for microfossils in the rock's interior. Scully correctly reminds Dr. Sacks that many scientists doubt that ALH84001 contains any fossils of ancient Martian life.

There are several reasons why a growing number of scientists are skeptical about fossils in ALH84001. To examine the meteorite, NASA scientists used the type of electron microscope that requires coating the subject with metal. Some geologists who specialize in the study of meteorites believe that the metal coating itself created the tiny objects that look like fossils. Other scientists believe that the fossils are just irregularities in the surface of minerals embedded in the rock. To counter these arguments, a French group recently used an advanced microscope called an environmental scanning electron microscope to examine a Tunisian meteorite called Tatahouine. Unlike conventional electron microscopes, this new device does not require coating the sample with metal. Looking at the Tatahouine meteorite in its natural form, they found objects that looked identical to the fossils in ALH84001. The Murchison meteorite from Australia also contains mushroom-shaped bodies that look like microfossils.

One of the arguments used to bolster the claims for Martian life is the striking resemblance of the tiny fossils to mysterious dwarf bacteria known as nanobacteria. Robert Folk of the University of Texas believes that nanobacteria are not only a major constituent of rocks and sediments buried deep below the surface of our planet, but are also found in tap water and tooth enamel. The size of nanobacteria, about one thousandth the volume of normal bacteria, is a cause of concern for many scientists. A cell this size is only big enough to contain a few ribosomes, a few hundred proteins, and only eight genes' worth of DNA. This is far, far less than what is necessary to support even the most primitive of bacteria.[6] For nanobacteria to be real cells, and not artifacts of the investigation methods, then they must have a biochemistry completely at odds with the biochemistry of all other organisms on Earth. While this may be a common occurrence in science fiction, it is much less likely to be true in real life.

---

[6]Scientists estimate that at least 250 genes are needed for a bacteria cell to survive and produce enough ingredients to allow it to replicate.

Other aspects of the report of ancient life in ALH84001 are also being questioned. Not every scientist believes that carbonates + magnetite + PAHs = life. At the heart of the controversy is the origin of the carbonate globules. Carbonates can form from water and carbon dioxide at low, life-compatible temperatures. It is crucial for the theory of life on Mars that there be evidence of liquid water. But carbonates can also form from liquid carbon dioxide as long as there is no water and a temperature of about 800 degrees Fahrenheit. Ralph Harvey of Case Western University and Harry McSween from the University of Tennessee believe that an asteroid crashed into Mars a billion years ago, liquefying the Martian surface and its covering of carbon dioxide frost, resulting in the carbonates found in ALH84001. According to Harvey and McSween, carbonates in the Martian rock do not imply the presence of water and life but just the opposite— searing heat in the absence of water, conditions where life could not possibly have existed.[7]

[7]If you aren't familiar with the scientific process, you might think that such diametrically opposed opinions when looking at the same piece of evidence have no place in the precise world of science. Ralph Harvey put the arguments over ALH84001 into perspective in the following piece that he wrote for the Internet:

> It is a completely normal part of science for researchers to "try on" various theories and interpretations, and at this stage utterly natural that groups might hold to contradictory interpretations. Normally the public doesn't see science at this stage, and I'm disappointed that they did in this case, because some might draw the conclusion that scientists are morons who can't make up their minds. In fact, that's the power of science—we take our time, examine all the possibilities, try hard to learn which data to trust and which data to mistrust. In the end, no theory answers all the questions, but usually a theory develops that successfully interprets the majority of the data, and offers predictions that are later seen to be fulfilled. Perhaps the most important aspect of science is that there's always some data somewhere that goes against the favored theory. Scientists don't ever discard those findings, because even though they are "sand in the gears," they often prove to be very important pieces of the puzzle in their own right, clues to phenomena we just didn't understand when we collected the data.

The tenor of this statement may seem familiar. It is what Scully faces week after week as she tries to make scientific sense out of events that can be beyond what science can easily explain.

Dr. Sacks is anxious to enter the debate on life in meteorites, so he prepares the meteorite for electron microscopy. Slicing through a section of the rock, Dr. Sacks must have instantly realized that an electron microscope would be superfluous. Oily black particles splatter onto the doctor, then congeal into slimy, sluglike worms. Before Dr. Sacks has time to be euphoric about discovering living organisms within the meteorite, the worms enter his body and he is paralyzed.

An organism that starts out as thousands of individual cells (like the black cancer particles) that then join together to form a single, multicellular creature sounds like a nightmare made for science fiction—but it isn't. *Dictyostelium discoideum* (known to fans as "dicti") is an organism called a cellular slime mold. Dicti goes through phases of its life where it resembles an ameoba, followed by an animal, then a plant, and finally a fungus. As an amoeba, single dicti cells are content to live a solitary existence, moving short distances and munching on bacteria for food. When an amoeba is hungry and bacteria are scarce, it releases a chemical that acts as a homing beacon to other dicti cells in the vicinity. The cells gather together and arrange themselves to form a single, multicellular slug. Dicti slugs bear an uncanny resemblance to the larger *X-Files* black cancer worms.

When the environment becomes dry, dicti slugs stop moving and undergo a transformation into plantlike organisms. In place of the original slugs now stand rigid, immobile "plants" that resemble very small caramel apples on tiny sticks. Spores form inside the "apple" part and burst through to the outside when mature. These spores with their tough cell walls resemble fungi and feed on decaying plant material. Eventually, each spore gives rise to an individual amoeba and the dicti life cycle begins again. I vividly remember an inventive college professor giving me and two hundred other students petri plates with dicti ameoba cells. I took mine back to the dorm and watched spellbound as slugs mysteriously sprang up from microscopic cells and then turned into little balls and sticks. For years I thought that if only dicti were one hundred times larger with a taste for human flesh, it would make a horrific science fiction menace.

While dicti slugs are harmless, the same cannot be said of the black cancer organisms in "Tunguska." Scully finds a colony of black vermiform (wormlike) organisms attached to Dr. Sacks's pineal gland, located just north of the hypothalamus in the brain. The pineal gland was historically viewed as the "seat of the soul" for its role in neurological and psychiatric disorders. One might think that the presence of vermiform creatures nesting in the pineal gland and paralysis imply cause-and-effect, but one would be wrong. Lab rats seem to live quite happily with their pineal glands removed. One dead giveaway that the black cancer creatures are cohabiting a human body is a black slimy film that appears over the eyes of the host. Since the pineal gland is directly connected by nerves to the eyes, the black film could be disrupting the perception of light by the host's pineal gland, thereby affecting the behavior of the host. In the *X-Files* episode "Piper Maru," infection by the same alien organism led to strange behaviors on the part of the human hosts, almost as if they were possessed by an occupying force.

Another major function of the pineal gland is to ensure that mammals without access to desktop calendars give birth during spring and summer, when the weather is nice and food is abundant. To synchronize the best times for romantic flings, mammals can unconsciously sense the time of the year by how long the sun is out during the day. Daylight is perceived by the eye's retina, which is connected by nerve cells to the pineal gland. The gland secretes the chemical melatonin, which controls fertility in females and the sex drive in males. It's not just rats and mice that obey their pineal glands and give birth during nice weather. Statistics for human births also indicate that more babies are born in spring and summer than in fall and winter. Unfortunately, Dr. Sacks is killed before Scully can complete her analysis on the connection between the pineal gland and the strange worms.

EXT. RURAL AREA (TUNGUSKA, RUSSIA)—DAY

KRYCEK

Tell me what we're doing here.

Mulder stops digging now, too. Looks at Krycek. There's a moment where we wonder if he might not rabbit punch him again. Then:

MULDER

June 30, 1908. Tungus tribesmen and Russian fur traders

looked up into the southeastern Siberian sky and saw a fireball streaking to earth. When it hit the atmosphere it created a series of cataclysmic explosions that are considered the largest cosmic event in the history of civilization. Two thousand times the force of the A-bomb dropped on Hiroshima.

Mulder continues to dig.

                         KRYCEK
What was it?

                         MULDER
It's been speculated that it was a piece of a comet or an asteroid, even a piece of anti-matter. The power of the blast felled trees in a radial pattern over an area of two thousand kilometers. But no real definitive evidence has ever been found to satisfy an explanation. Of what it was.

Mulder has moved enough earth to slip under the fence now. As Krycek watches him crawl to the other side.

                         MULDER
I think someone found that evidence. And I think the explanation might be something no one ever dreamed.

                                              —"Tunguska"

Meanwhile, Mulder travels to Russia to search for the origin of the meteorite containing the alien organism. After trekking through the remote wilderness of Siberia, he finds workers unearthing a huge meteorite in the Tunguska region. Mulder would most certainly understand the significance of finding a meteorite in this location. Tunguska is the site of the largest natural explosion in the history of civilization.

At 7:17 A.M. on June 30, 1908, local tribesmen in the trading post of Vanivara in Siberia saw a ball of fire blazing across the cloudless sky followed by a deafening explosion. Whatever occurred four to five miles above the rolling hills and forests of the Tunguska river region released a blast of energy straight downward that was two thousand times greater than the atomic bomb that devastated Hiroshima. Seismographs from around the world picked up pressure waves from an event so catastrophic that the waves circled the Earth twice. Trees, more than 60 million, toppled over in an area the size of Rhode Island. Fiery lights flickered for days afterward, illuminating midnight skies all over Europe and Asia. The lights were so bright that citizens of London thought that their city was on fire. Although the event probably devastated reindeer herds in the isolated region, only two people died. Tunguska is so remote that it took nineteen years for the first Russian expedition to reach the devastated site. Even today, visiting the site means a fifty-mile hike if a convenient helicopter is not available.

Roy Gallant, director of the Southworth Planetarium at the University of Maine, was the first American to visit Tunguska. Roy told me that it takes an experienced eye to see any remaining traces of the catastrophe. Directly beneath where the explosion occurred stands a forest of telephone poles—sixty-foot trees stripped of their branches that are stark reminders of the downward force of the blast. Remains of other trees three to ten miles away lie on the forest floor in a radial pattern pointing outward from the center of the cataclysm. Otherwise, there are no craters, no signs of further disruption to a landscape that must have once looked like ground zero after an atomic blast.

There are two mainstream explanations for what caused the explosion in Tunguska. One group of scientists led by Chris Chyba, Peter Thomas, and Kevin Zahnle are convinced that minute rock fragments buried in trees point to the explosion of a stony meteor measuring fifty to sixty yards in diameter. Others, including Roy Gallant, believe that only a comet weighing in at over 100,000 tons could explain the mysterious lights that preceded the event by several days. Mulder would probably favor one of the more unusual explanations—a nuclear explosion from a crashed UFO or passage of a black hole through the Earth. However, a little scientific investigation by Scully would fail to find a trace of radioactive material in the region. She also would conclude that a black hole would have exited through the fishing grounds of Newfoundland, Canada. Why no fisherman bothered reporting the deafening explosion of a black hole shooting out of the ocean or the immense tidal wave that followed are questions that somehow haven't deterred the black hole enthusiasts. According to the *X-Files* episode "Tunguska," if scientists had dug in the right place or at the correct depth they would have found the real culprit—a huge black meteorite filled with the alien black cancer organism.

In "Tunguska" and the earlier episode "Ice," meteors are the delivery systems for alien hitchhikers that are able to invade a human host. This brings up several very interesting questions. What are the chances that an alien microbe could survive a trip through space on a vehicle that lacks such basic amenities as an atmosphere, temperature control, sunscreen, refreshments, and brakes? And providing that an organism could survive the rather abrupt explosion or crash landing at the end of the flight, what are the chances that it could become a parasite of Earth's creatures?

Let's first consider whether a microbe could survive an interstellar cruise on a rock. You might be surprised to learn that respected scientists have been asking this rather unusual question for nearly 100 years. At the turn of the twentieth century, a Swedish electrolyte chemist named Svante Arrhenius wondered if bacteria could survive in outer space. His curiosity on this subject must have earned him many puzzled looks and raised eyebrows from colleagues. Arrhenius's delving into the realm of science fiction was spurred by a discovery that seems at first glance to be unrelated to space-faring microbes—he found that light can exert a force on objects, if the objects are very, very small. Arrhenius observed that clouds of tiny particles blasted off the icy core of a comet are swept back away from the sun like sails in a solar breeze. That is, light waves exert enough force to push small ice particles, like wind pushing the leaves of a tree. As a physical chemist, Arrhenius determined mathematically that objects would need to be between the size of a bacterial spore and a tiny dust particle to be moved by light. Therefore, Arrhenius speculated, if bacteria could survive in space, they could sail between the stars powered by gentle solar breezes.

At first glance, bacteria in space seems like a fantasy worthy of Mulder's imagination. The hardships bacteria would need to overcome seem overwhelming. If the microbes don't die of asphyxiation or dehydration then the intense cold of near absolute zero should turn their cellular liquid into jagged ice spears. Microbes would have as much chance surviving floating freely in space as they would have surviving being stranded on the moon.

But bacteria *can* survive being stranded on the moon. In 1967, the unmanned lunar lander *Surveyor 3* set down gently on the surface of the moon and sent back detailed pictures of a future manned landing site. The cameras of *Surveyor 3* remained on the moon for two and a half years before being picked up by *Apollo 12* astronauts and returned to Earth. NASA scientists were astonished to find that tiny bacteria had stowed away in the camera before *Surveyor 3* was launched from Earth. These bacteria survived their odyssey with no water and after repeated cycles of being frozen and thawed, courtesy of the moon's huge monthly temperature swings. Pete Conrad, commander of the *Apollo 12* mission, considered the survival of the wayward bacteria to be the most significant finding of the mission. In retrospect, Pete might not have been surprised at the hardiness of bacteria had he known what Arrhenius knew—bacterial spores can withstand desiccation in a powerful vacuum and still grow and reproduce afterward.

Surviving in a frigid vacuum is only one hurdle to overcome. If bacteria are to survive the rigors of unprotected space, they must live through constant buffeting by deadly cosmic rays and ultraviolet radiation. These high-energy emissions from the sun make mincemeat of DNA and other essential molecules by blasting apart chemical bonds. Bacteria should have as much chance surviving the intense radiation of interstellar space as they would have surviving in extremely radioactive water surrounding the core of a nuclear power plant.

You guessed it. There are bacteria living in radioactive water inside nuclear power plants. The appropriately named *Micrococcus radiophilus* manages to exist in conditions that would instantly kill a human and even turn glass brown and crumbly. Bacteria such as *Micrococcus radiophilus* have evolved several ingenious strategies for withstanding exposure to radiation. The first line of defense is a shell made of sugar and protein that surrounds a bacterium's fragile plasma membrane. The protective nature of the shell can be increased with a dark pigment that acts like sunscreen and is used by bacteria that live in the upper atmosphere. Bacteria also have a task force of enzymes that constantly monitor and repair errors and damage in their DNA. Why, some scientists argue, have bacteria evolved such abilities if not to cruise the heavens?

So it is not outside the realm of extreme possibility that cells might survive on meteors such as the one portrayed in "Tunguska." Proponents of this theory go on to speculate that life on Earth may have originated from microbes landing from outer space. This theory of the origin of life was first voiced by Arrhenius, who called the idea *panspermia.* Believers in panspermia contend that life did not develop on Earth but was seeded by bacteria from outer space. Panspermia, proponents suggest, is the only answer to the question of how life could have originated from nonlife in less than—perhaps considerably less than—750 million years. Like Mulder, panspermia proponents suggest that aliens might be invading all the time as stowaways on comets, meteoroids, or space dust. Believers in panspermia are not just science fiction junkies. Francis Crick (Nobel laureate and co-discoverer of the structure of DNA) supports the theory of panspermia as does noted English astronomer Sir Fred Hoyle. Even Dana Scully, in "Biogenesis," the last episode of the sixth season, concedes that panspermia is a plausible hypothesis. Because it is accepted that bacteria can survive under conditions once thought impossible, debates on the possibility of panspermia now appear in major scientific magazines.

How much life might be cruising the galaxy, hitching rides on comets or sailing on dust? Quite a bit, according to Fred Hoyle. Hoyle, with his colleague Chandra Wickramasinghe, made this highly controversial assertion in 1979 while studying a curious cosmic phenomenon—galactic dust clouds. First a little history: In the 1920s, the spectrum of light reflected by galactic dust clouds indicated that water was a major dust cloud component. By the 1950s, more sensitive instruments revealed a host of other substances. But what? Hoyle and Wickramasinghe suspected that dust clouds contained a mixture of water and some carbon compounds, possibly graphite (like pencil lead). They tried various mixtures of ice and graphite in the laboratory to see if they could re-create the same peaks and valleys of the light spectrum reflecting off the dust clouds, without success. Another baffling mystery was that only nearly hollow particles of about one micrometer in size could explain a discrepancy between the light diffracted by the clouds and their apparent mass. The puzzle seemed impossible to solve.

For years, Hoyle and Wickramasinghe pondered their riddle. What was mostly water and carbon, with a dash of other common elements thrown in, hollow, and about a micrometer in size? In 1979, the answer hit them: bacteria. In the laboratory, bacteria fit the spectrum of light bouncing off the dust clouds perfectly. Needless to say, the declaration of Hoyle and Wickramasinghe that clouds of bacteria are sailing through the galaxy provoked an outpouring of criticism—but no one has yet to provide another answer for the riddle of the galactic dust clouds.

So there are structures resembling fossils of microbes in meteorites and evidence that bacteria could survive a trip through space. However, other interpretations of the evidence are also possible. Deep down, like Scully, I will only be convinced when a sample is retrieved off a comet or planet and life is found that could not possibly have originated from Earthly contamination.

Missions are now being planned to do precisely that—to retrieve samples from other planets. One of the most likely places in the solar system where life may exist is Europa, Jupiter's smallest moon. Europa is one of the most brilliant objects in the solar system. Its flat surface reflects light off a shimmering layer of ice. The *Galileo* space probe recently traveled so close to the surface that cameras could have picked up a vacation cottage. While such obvious signs of life are unlikely given the frigid -260 degree Fahrenheit surface temperature, pictures provided tantalizing evidence that conditions below the surface may be suitable for living creatures. Photographs show striation patterns indicating that the ice sheet had been broken and rebroken many times—signs that a turbulent liquid ocean once flowed and might still flow beneath the ice. Other experiments on board *Galileo* indicate that both Europa and its sister moon Callisto have interiors that are able to conduct electricity. The most likely explanation? The electrical conductor is a substance contained in liquid water. The very strong possibility of liquid water means that the internal temperature of the moons are sufficiently warm to create an ocean of water.

The likeliest heat source so far from the sun is the internal friction of the moon itself moving over a molten core, the same tectonic motion on Earth that frees heat vents in the ocean floor. Since on Earth thermophilic (heat-loving) bacteria thrive around such vents at temperatures above the boiling point of water—more evidence, if such were needed, for the ubiquity of life on, in, and under the earth's surface—similar life on Europa appears more and more probable. There is even an atmosphere, with a smattering of oxygen molecules.

Missions are already being planned that will answer the question of whether life is present below the surface of Jupiter's glowing moon. Determining whether the ocean still exists will require landing a probe on the surface, a mission that NASA is studying for the early twenty-first century. The search for life, however, requires more creativity. One plan is to send a pencil-shaped robot that will plunge itself into the ice and deliver an aquabot that will search and report back. Another idea is to send a spacecraft that will orbit Europa and drop a boron-nitride bomb on the ice. The explosion will send up a plume of chunks and chips that can be grabbed when the craft flies through it.

If life is found below the surface of Europa or Mars, would it be similar to life on our planet? Would it share the same biochemistry as ours? The answer depends on the validity of theories like panspermia. If there was a single genesis of life in our solar system, then cells on other planets may share common characteristics with life on Earth, such as DNA for the genetic material and being surrounded by plasma membranes. Some divergence from terrestrial life would certainly be found since life on other planets would be shaped by different environmental conditions. However, bacteria that have lived below our own ground for hundreds of millions of years survive in environments completely unlike the habitats of bacteria above ground, yet these deep dwellers are remarkably similar to their aboveground cousins. A few simple experiments such as those performed by Dr. Carpenter on the contents of the Erlenmeyer flask should be sufficient to establish whether a relationship exists between any cells from Europa and cells from Earth.

There is a possibility that if alien bacteria shared a single genesis with our bacteria, they could have a harmful effect on human physiology. For this reason, stringent containment measures are planned for any rocks returned to Earth. However, the odds should not be any greater than for random Earth bacteria to be pathogenic on humans. Of the countless species of bacteria on our planet, very few are harmful.

INT. ARCTIC COMPOUND—NIGHT
MICROSCOPE MATTE

If a single celled creature could appear mean and vicious, this is it. Protective spikes encircle the undulating membrane. A whip-like flagellum propels the creature in an excited state.

> MURPHY (O.S.)
> Tell me that's not a foreign object.

WIDER

Mulder pulls away. Scully looks into the tube. What she sees seems to take her breath away. Scully looks to Mulder.

> SCULLY
> The same thing is in Richter's blood.

Everyone tenses. She gestures to her work area. Mulder moves to the other microscope. Bear watches intently, nervous, as Mulder peers into the tube.

MICROSCOPE MATTE

Indeed, the same kind of creature is moving about.

WIDER

Mulder pulls up from the scope. Scully turns to the others.

> SCULLY
> That single celled organism could be the larval stage of a larger animal.

> HODGE
> That's a big leap, Scully.

SCULLY

The evidence is there.

—"Ice"

If alien life-forms were unrelated to terrestrial life, could they still be harmful? This would depend on the biochemistry of the alien organism (and whether they have ray guns). For example, let's speculate about the biochemistry of the weaponless alien worm in the *X-Files* episode "Ice." The worm was discovered in an ice-core sample that was retrieved from deep below the ice of northern Alaska near a meteorite. The ice surrounding the worm contains high levels of ammonia. Mulder speculates that the alien worm originated on a planet with an ammonium atmosphere.

There are many qualities of ammonia that make it ripe for extraterrestrial speculation. Such speculation usually begins with an awareness of the importance of liquid water to life on Earth. Water's structural properties (and omnipresence) are so important to every form of earth-based life that most scientists believe that extraterrestrial life is likely to be found only on planets where water exists in its liquid form. As this represents a fairly narrow range of temperatures—32 degrees to 212 degrees F—the number of possible planets is similarly restricted . . . that is, unless a substance like water can be found. Consider, therefore, ammonia.

Ammonia consists of a nitrogen atom bonded with three hydrogen atoms. Water is an oxygen atom bonded with two hydrogens. Both molecules are polar. This means that the electrons shared between atoms forming a chemical bond are not shared equally. The electrons are more strongly attracted to the nitrogen or the oxygen and therefore spend less time with the bonded hydrogens. Since electrons have a negative charge, by spending more time with the oxygen or nitrogen than the hydrogens, the oxygen and nitrogen have a more negative charge and the hydrogens are more positively charged. Since negative- and positive-charged atoms attract, the more positive hydrogens are attracted to atoms on other molecules that have a more negative charge. The polar property of water—the "universal solvent"—is why it can dissolve proteins, salts, and sugars. Since ammonia is also polar, the essential position of water as the liquid of life on Earth would be filled by ammonia on an ammonium world.

On our temperate planet, ammonia evaporates at room temperature, which is what accounts for its strong odor. If it were much colder and the atmosphere was soaked with ammonia, ammonia would form pools of liquid just like water. Imagine then an ammonia planet: freezing cold with oceans of ammonia. Mulder is correct in stating that if the worm's planet had an ammonium atmosphere, it would have to be a frigid place. Based simply on outside temperature, the worm would feel right at home in the ice of northern Alaska.

Could life develop on a planet with an ammonium atmosphere? Where water is scarce and liquid ammonia is abundant? Possibly. Imagine a world where the primordial soup is mainly ammonia with a dash of carbon dioxide, phosphate, and some metals. Substitute water for ammonia, and conditions might be similar to Earth's during the dawn of life. The ammonia world would need an energy source, say UV light radiating from a nearby star. Our virgin ammonia planet would be a beautiful place. Yellowish-orange oceans of ammonia shimmering beneath a thick, gaseous atmosphere of ammonia, nitrogen, and carbon dioxide. Lightning storms would deliver torrents of ammonia rain to the surface.

Within a few billion years, the first life-forms could evolve on the planet: tiny, single-cell microbes capable of using the sun's energy to build the complex molecules of life—microbes capable of photosynthesis. On our water world, the sun's energy powers the photosynthetic conversion of water and carbon dioxide into oxygen and sugars. On the ammonium world, photosynthetic microbes could use their sun's energy to convert ammonia into nitrate, by replacing the hydrogens bonded to the nitrogen in ammonia with oxygens. At the same time carbon dioxide would be converted into sugars. New microbes could evolve that feed on the photosynthetic microbes, which eventually could lead to a type of eukaryotic cell.

Creatures of an ammonium world would probably be carbon based. Advanced organisms would eat food rich in nitrate and carbohydrates (sugars). They would breathe in nitrogen and exhale carbon dioxide. The alien worms from an ammonium planet would probably drool at the sight of freshly fertilized plants, since plants are filled with carbohydrates and fertilizer contains nitrates. If the alien worms proved unfriendly, guns and bombs would not be required, just a garden hose. Water would be as noxious a poison to them as ammonia is to us.

On our water-based world, ammonia is a household product handy for killing bacteria. Ammonia is toxic because, just like water, it passes through the plasma membranes of cells. When ammonia gets into bacteria and fungal cells, it acts like drain cleaner, causing instant death. Water should have the same effect on ammonia-based creatures. The worms from the episode "Ice," if they originated on an ammonium-soaked world, would be poisoned by water long before they ever reached the brain. No creature from an ammonia-based planet could ever be a parasite of a water-based organism.

Whether or not you believe in alien hitchhikers, three hundred tons of extraterrestrial organic matter fall to Earth every year courtesy of comets, meteors, and dust. Four billion years ago, when life on Earth was first appearing, there were hundreds or thousands of times as many comets and meteors on collision courses with Earth. William Irvine, a University of Massachusetts astronomer, estimates that enough comets, meteorites, and dust have fallen over the years of Earth's existence to provide all the raw organic materials present on the planet. Life may or may not have originated on our planet, but our molecules are nevertheless the molecules of the universe.

Sea Urchins in the Arctic

INT. MULDER'S APARTMENT—EARLY MORNING

> SCULLY
>
> I have come here today, four years later, to report to you on the illegitimacy of Agent Mulder's work. That it is my scientific opinion he became, through the course of these years, a victim of his own false hopes, and of his belief in the biggest of lies.

According to Fox Mulder, proof of the existence of extraterrestrial life would be the greatest discovery in the history of science. It is the holy grail of his life, the search for his sister, kidnapped during a terrifying evening of bright lights when they were both children. In "Gethsemane," the final episode of the fourth season, Mulder be-

lieves that he has finally found the elusive proof. A Canadian geo-
detic survey team, which specializes in mapping land while taking
into account the curvature of the earth's surface, finds a body frozen
in the wall of an ice cave. It looks like the quintessential alien: small
stature, large bald head, wide oval eyes, gray wrinkled skin, and in-
determinate sex . . . another of the "grays," an alien body that fits the
description given by numerous "abductees" and Roswell, New Mexico, "eyewitnesses." If it is a hoax, it is a deadly serious one. When an
excited Mulder finally reaches the remote site, he is greeted by the
bullet-riddled bodies of the survey team. Fortunately, the frozen
alien was hidden by the one surviving member. After returning to
civilization, an autopsy of the body reveals masses of stringy white
tissue not found in normal humans.

Mulder believes that the alien corpse is authentic while Scully is
skeptical. Besides the body, the only hard evidence is in the form of ice-
core samples taken from the cave. If the ice-core samples prove to be at
least two hundred years old, the last time that the cave was not covered
with ice, then a body near the ice samples should also date from the
same time. Since people in the eighteenth century probably weren't
squirreling away fake aliens in the Arctic, the date that the body was
frozen would be an important clue in establishing the truth.

At Mulder's request, Scully takes the ice samples to Dr.
Vitagliano at the Exobiology Laboratory, Goddard Space Center. At
first glance, the samples are just what would be expected for ice that
has not melted for hundreds of years—layers of sedimentation with
the surface layer containing hydrocarbon pollutants that not even
the Arctic can escape. No surprises, until Dr. Vitagliano analyzes the
ice sample more closely. Using a light microscope, he sees that within
the ice matrix are cells. When Dr. Vitagliano tells Scully that cells are
in the ice, she asks if the cells are from a plant or animal. His answer
is cryptic. The cells cannot be classified as plant or animal.

Scully is surprised by Dr. Vitagliano's response because she
knows that plant and animal cells are easy to differentiate. Only
plant cells have a sturdy rectangular wall composed of proteins and
sugars that surrounds the plasma membrane. Plant cells also have
chloroplasts, the little cell-like organelles that help make sugars from
sunlight, water, and carbon dioxide. Taking up much of the interior
space in some plant cells is an enormous bubble called a vacuole, the

storage bin and waste pit of the cell. Finally, plant cells do not have little cubicles called lysosomes, which are used by animal cells for storing enzymes that would be detrimental if floating around free. Besides these differences, which are easily observable using a light microscope, plant and animal cells are remarkably similar. Both have nuclei containing the DNA genome and a main compartment within which float membranes, ribosomes, mitochondria, and other constituents.

For the cells to be unclassifiable, they must have a mixture of plant and animal features. Since such cells do not exist in nature and are not shown on camera, let's be imaginative and speculate. Dr. Vitagliano may be looking at cells that lack a cell wall and contain lysosomes like animal cells while having organelles that look like chloroplasts and a large vacuole. Alternatively, the cells could have a cell wall but be completely animal-like within the plasma membrane. When Chris Carter was writing this episode, he phoned to ask what he should call cells that are a mixture of two organisms. I told Chris that these are known as hybrid cells, or chimeras. The word *chimera* dates back to a creature from Greek mythology, a fire-breathing monster with a lion's head, a goat's body, and a serpent's tail. When I received the script to check it for accuracy, I was amused to see that Chris had used both terms, calling the cells "chimerical hybrids." Chris and I had several conversations about hybrid cells over the years, beginning with the episode "The Host," which contained a charming chimeric organism known as flukeman.

Chris was intrigued to learn that any two cells can be fused together providing that they both have plasma membranes at their outermost surfaces. For plant cells, this means removing the protein-sugar wall that encloses the plasma membrane. The reason why any two cells can be fused is that proteins and lipids move freely within a membrane, making the membrane a highly fluid structure. Bring two cells together, and the membranes flow into each other, much like two drops of liquid mercury can coalesce to become a single large drop. Hybrid cells can result from the fusion of different types of cells from the same organism, say muscle and liver cells, or from the fusion of cells from two different organisms, like humans and aliens. On my suggestion, Chris made the fusion of fluke and human cells one of the possibilities for how flukeman was generated.

Science defines hybrid organisms as the progeny of parents that normally would choose not to intermingle, like man and fluke. Distinct species of organisms arise because members either cannot or will not breed with other organisms. While plants can form fertile hybrids between species, even between genera—some highly decorative orchids are derived from as many as four different genera—hybrid animals are rare and sterile. Organism development is a complicated process and both genomes of a hybrid organism must work closely together, which is not usually possible. Mules are the best-known animal hybrids, the natural offspring of male donkeys and female horses. Beefalo, a new hybrid animal whose meat is available in my supermarket, is produced from crossing cows and buffalo. Creatures like flukeman and human-animal hybrids in fictional stories like H. G. Wells's *The Island of Dr. Moreau* are pure fantasy.

While Scully and Dr. Vitagliano are puzzling over the strange hybrid cells, Mulder's belief in the authenticity of the alien body is shaken when he is told by a Department of Defense employee named Kritschgau that he has been fooled by an elaborate hoax. The body, Kritschgau tells him, is really an artificial creation from the genetic manipulation of chimeric cells. What Kritschgau seems to be stating is that the alien is merely a strange-looking hybrid, cleverly concocted in a laboratory from cells of two natural terrestrial organisms. This would make the alien simply a modern science version of Piltdown Man.

Piltdown Man is the most famous fraud in the history of science. The hoax began in 1912 when amateur archaeologist Charles Dawson discovered fragments of a skull in the Piltdown quarry in Sussex, England. The skull was a tremendous boost for English anthropologists, who felt left out after Continental scientists discovered Neanderthal, Cro-Magnon, and Java Man remains. Piltdown Man outshone all other finds. It was the "missing link," a humanlike skull with elephantlike teeth and an apelike jaw. And that's what it was. A human skull, elephant molars, and an orangutan jaw.

The perpetrators of the hoax were very clever. The teeth were filed to fit the jaw. The skull was broken to obscure the nonexistent connection with the jaw. The bones were boiled and treated with an iron solution to make them look old. When scientists voiced doubts that the jawbone and skull seemed to be from different animals that coincidentally died in the same location, a new skull and jaw were miraculously found. One such coincidence, maybe. Two? It had to be from the same creature. For forty years, anthropologists puzzled over Piltdown Man, the only anomaly in the increasingly fleshed-out theory on the evolution of man. The hoax survived until a new fluorine dating test in 1949 established that the skull was merely six hundred years old and the jaw was twentieth-century ape. Still, it took another four years before a paper was published establishing the nature of the hoax.

Just like Mulder can be fooled by an alien hoax because he so desperately wants to believe, British anthropologists accepted the lie because it was precisely what they wanted to find. At the time, the leading theory of man's evolution was that the braincase would develop before the jaw, leading to a humanlike skull and a simian jaw. Piltdown Man fit the theory perfectly. The British were also pleased because finding Piltdown Man in England meant that the first intelligent humans were, of course, British. Another reason why the Piltdown Man fraud was so easily perpetrated was that none of the principals seemed to have anything to gain. No one became rich over the finding, although reputations were certainly enhanced. Amazing scientific discoveries made by reputable people don't usually elicit accusations of fraud. What the anthropologists needed at the time was an objective scientist like Scully with access to the bones. Someone who would have taken better X rays, tested both the skull and the jaw for similar organic matter, and used a microscope on the teeth to discover the very obvious signs of tampering.

The perpetrators of the Piltdown Man hoax remain a mystery. Was it Charles Dawson, the finder of the "fossils" and now known to be a notorious forger and plagiarizer? Or perhaps it was Sir Arthur Woodward, the Keeper of Geology of the British Museum's Natural History Department who had access to bone collections that could be used to fake the remains. Teilhard de Chardin, before he became a prominent theologian, accompanied Dawson and Woodward into the field, and has since been accused—if that's the correct word—by Harvard paleontologist Stephen Jay Gould. Even Sir Arthur Conan Doyle, famous writer of the Sherlock Holmes mysteries, has come under suspicion since he lived in the area.

When Mulder asks Kritschgau why people would want to perpetrate an alien hoax, he replies that the lie is there to divert attention from greater lies. Mulder responds that science today is quite capable of testing the body to discover the truth. But before testing can begin, the body is stolen. The question remains unanswered: Is it real, or is it an elaborate hoax?

If the body had not been stolen, Kritschgau's information that the alien was just a chimera of known terrestrial species could easily have been tested. Over the years, Mulder and Scully have frequently relied on DNA analysis to help determine if strange-looking corpses were man, ape, or alien. DNA can be used as a precise fingerprint because every organism on Earth, except for identical twins and clones, has its own unique genome. The precise order of the 6 billion nucleotides in your genome is a mixture of your parents' genomes. If you were to compare your nucleotide order with that of any other person's, they would differ by only about 0.1 percent, which is still a considerable 6 million nucleotides.[8] If your DNA is only 98.5 percent the same as the DNA of humans, you would be swinging through the trees with your fellow chimps.

If someone commits a crime and leaves a trace of themselves behind—a wisp of hair, a drop of blood, a flake of skin—then DNA in these samples can be matched to the perpetrator's DNA. The current ability to match a sample of DNA to the person it came from requires the identification of regions of the human genome that tend to vary among people, the so-called variable regions. The chances that two people will have exactly the same DNA sequence in ten different variable regions is as low as one in several billion. Variable regions of DNA, most of which have no known function, tend to be located in between genes. The reason for variability in nonfunctional DNA regions is simple. Nucleotide changes, also called mutations, can accumulate in nonfunctional parts of the genome over the millions of years of human existence without harm. In the same way, bad sectors that accumulate in unused portions of a computer's hard drive are substantially less damaging than bad sectors in the operating sys-

---

[8] As psychologist and philosopher William James put it, "There is very little difference between one man and another, but what difference there is, is very important."

tem of the computer. Genes have to specify the building of a precise protein. Mutations in the DNA of a gene can seriously affect the composition and function of the specified protein. Acquire a new mutation in an important gene and the individual might never develop normally enough to be born.

The most common test to match DNA samples is called restriction fragment length polymorphism, or RFLP. When Mulder and Scully talk about the need to conduct DNA analysis on a sample of human blood or tissue, this is usually what they mean. The RFLP test was used in the O. J. Simpson murder trial to determine whose blood was on objects of evidence. It was also used to identify the remains of the Romanovs, the Russian imperial family murdered in 1917, whose bones lay hidden for seventy years.

The RFLP test makes use of natural bacterial enzymes called restriction enzymes, which recognize small sequence patterns in DNA. For example, the enzyme called EcoRI recognizes the nucleotide sequence GAATTC (guanine-adenine-adenine-thymine-thymine-cytosine). The enzyme will snip the DNA double helix wherever this order of nucleotides appears. If GAATTC is found once in a piece of DNA, then the enzyme will cut the DNA at its location, resulting in two pieces of DNA. In a DNA chain that is millions of nucleotides long, the sequence GAATTC will appear many times and the DNA will be cut into many pieces after treatment with the enzyme.

In a variable region of the DNA, not all people will have the same order of nucleotides. If the sequence GAATTC appears in a variable region of the DNA, then the DNA of some people might be slightly different. Instead of GAATTC, some people might have TAATTC. The DNA of people with TAATTC will no longer be cut at this location by the EcoRI enzyme. Since the fragments of DNA after enzyme treatment are separated according to their size, people will have different sizes and numbers of fragments depending on whether the enzymes cut or didn't cut the DNA at a particular place. To perfectly match a human DNA sample with a particular person, a variety of different restriction enzymes that each recognize a different short sequence of DNA are used to examine a number of different variable regions of the genome.

RFLP analysis is useful if you need to identify a specific person. To determine if tissue is human, ape, alien, or some weird chimera, it is easier to simply determine the order of nucleotides in a particular region of DNA. The nucleotide order in the ribosomal DNA sequence has already been determined for thousands of organisms. By sequencing this gene from the DNA of an unknown creature, you can immediately identify its origin or determine if it has a known close relative. Just as (the late) Dr. Anne Carpenter sequenced the ribosomal DNA gene to determine the nature of the unknown bacteria in the episode "The Erlenmeyer Flask," sequencing the ribosomal DNA gene from a sample of the "fake" alien's tissue would tell a scientist the organism from which the tissue came.

Mulder tells Scully that despite Kritschgau's tale of alien hoaxes, he is still inclined to believe that the alien body is genuine. Scully informs Mulder that the chimeric cells found in the ice matrix support Kritschgau's story of how the fake alien was engineered in the lab. Mulder argues that the chimeric cells could be extraterrestrial. Scientifically, Mulder has a point. Kritschgau's explanation that chimeric cells were used to construct the alien indicates a technology that is well beyond what science can accomplish on this planet.

In the *X-Files* episode "Redux," the sequel to "Gethsemane," Dr. Vitagliano continues his tests on the chimeric cells from the ice-core sample to determine if they are capable of cell division. In other words, he wants to determine if the cells are still living. For real hybrid cells to be capable of cell division, their parent cells must be closely related, like mice and humans. Chris Carter, who wrote the episode, wanted to know how Dr. Vitagliano could discover that not only were the chimeric cells alive, but that they were developing into an organism. I told Chris that Dr. Vitagliano should place the cells in a nutrient-rich solution. If these were normal animal cells, he would use a pink-colored liquid broth enriched with fetal bovine serum. Dr. Vitagliano tells Scully that when he placed the cells in the media, the cells began to divide.

Since Scully would know that media containing baby cow blood would not please even a carnivorous plant, she assumes that the cells must be from an animal. Dr. Vitagliano still maintains that the cells cannot be classified. Scully is confused, and tells him that the cells were able to complete mitotic cell division in the media.

Eukaryotic cells can divide either by mitotic cell division (mitosis) or meiotic cell division (meiosis). Whether the cells are chimeras from the Arctic or parts of your lung, nearly all will divide by mitosis. In mitosis, one cell becomes two, and the two daughter cells are perfect copies of the parent cell. Many of the 100 trillion hardworking cells in your body have very short life spans. One cell dividing to give two is the only way to replace the billions of cells in your body that die every day so that you can live. Skin cells last only between one and thirty-four days and the cells lining your stomach give out after only two days. Not happy with your current liver? If the cells are healthy and able to divide, you'll have a new liver every five hundred days. Some cells in your body don't divide and cannot be replaced. Nerve cells, such as your brain cells, and heart muscle cells are designed to last a lifetime, but the manufacturer's warranty doesn't cover self-destructive activities.

Only cells called germ cells, which give rise to egg and sperm sex cells, divide by the process called meiosis. In meiosis, daughter cells end up with exactly half the DNA of the parent germ cell. This way, when egg and sperm cells join together, they each contribute half of the DNA to the composite cell, called a zygote. From a single cell, the zygote starts mitotic cell division and develops into a complete organism.

Since the ice-core sample cells don't appear to be germ cells, they must divide by mitosis, as Scully suggests. Scully expects the cells to divide like average animal cells would if bathed in nutrient media—the two daughter cells separate completely from each other after the division. Instead, she is shocked to learn from Dr. Vitagliano that when the cells started dividing, they began to go through the stages of morula, blastula, and gastrula.

I explained to Chris how Dr. Vitagliano could tell that the cells were developing into a complex life-form. In many animals, after the sperm cell fertilizes the egg cell to form a zygote, the zygote starts mitosis and the resulting glob of cells goes through defined stages of development that can be easily identified using a light microscope. Morula, blastula, and gastrula are three of the early developmental stages. Only animal zygote cells have the complete genetic program and activated cytoplasm necessary to proceed through the many stages of organism development. In plants, it's different. While normally a plant develops from a zygote, which is produced when pollen fertilizes an ovule, I had talked to Chris about how I had taken ordinary plant cells and tricked them into believing that they were really "zygotes." The cells proceeded to go through the many stages of plant embryo development. This type of development, which does not begin with a real zygote, is called somatic development. The term "somatic" refers to the type of cell undergoing development. All cells of an organism that are not germ cells are somatic cells.

Scully and Dr. Vitagliano agree that the cells have begun somatic development, which will eventually result in a life-form. For this scene, Chris needed a real live alien organism. Since one was not readily available, he asked for my help. What could he use that would look alien and wouldn't be easily recognized by the average *X-Files* fan? Tough question. After mulling the possibilities with my scientist husband, Cliff, for several hours, we agreed on our alien—a pluteus, one of the stages in the development of the common sea urchin. And I knew just where Chris could film a pluteus. One of the foremost labs studying sea urchin development was a stone's throw away from 20th Century Fox, the home of Chris's Ten Thirteen Production Company: Eric Davidson's lab at the California Institute of Technology.

When the audience is treated to a view through the microscope of the developing alien, they see the pluteus, a translucent, pulsing creature with several protrusions coming from one end and a pointed other end. Scully and Dr. Vitagliano are mesmerized by the creature, and with luck, so was the audience. I wondered at the time if any viewers would recognize that the alien creature was really a sea urchin. The next day, I received an answer. A faculty member friend told me that she watched the episode with her husband, a novice *X-Files* viewer and scientist who studies marine organisms. After the scene where the pluteus "alien" is shown, he turned to my friend and with a puzzled look on his face asked, "What are sea urchins doing in the Arctic?"

In "Gethsemane" and "Redux," as well as "The Erlenmeyer Flask" and "Tunguska," aliens are not little green men or costumed monsters, but something far more sophisticated . . . and plausible. The aliens are cells. If we are fortunate enough to discover life on Mars or Europa, it is cells that will be brought back to Earth for study. As a scientist, I shiver with anticipation when I think about that hypothetical test tube in the containment room at Fort Detrick. Will cells be found? Will they prove that life in our solar system had a single genesis? And if not, what strange metabolism will these cells have? What type of genetic material? Will they have plasma membranes? Will they have the same twenty amino acids in their proteins as on Earth? My questions are endless. Aliens do not have to step out of spaceships saying, "I come in peace," or blow up the White House to be worthy of our attention. I agree with Mulder. Proof of the existence of extraterrestrial life would be the greatest discovery in the history of science.

It would make for pretty compelling television, too.

3

Mutants and Monsters

Introduction

INT. DR. POLLIDORI'S OFFICE—NIGHT

Dr. Pollidori turns back to the monitor, on which is now a grotesque blow up of a fruit fly head.

> DR. POLLIDORI
> Behold Proboscopedia.

> MULDER
> (looks closer, amazed)
> This fly has legs...

> SCULLY
> ...growing out of its mouth.

> MULDER
> (disturbed)
> Why would you want to do that?

> DR. POLLIDORI
> Because I can.

> —"The Post-Modern Prometheus"

A few years ago, I told Chris Carter about a mutant monster every bit as frightening as anything on *The X-Files*. I described its bulging red eyes, spindly neck, and hairy chest that sported three pairs of legs and a pair of wings. A yawn fest, you are thinking? Well, what about an extra pair of legs attached to its forehead or, better yet, coming out of its mouth? And best of all, this monster is no fabrication of a slightly deranged mind. This monster is very real.

Picture the following scene: FBI agent Fox Mulder is relaxing on the couch in his apartment munching on sunflower seeds. A large open window relieves the heat of a midsummer's day. Suddenly, the mutant monster flies in toward the couch, legs flailing from its mouth. Does Mulder jump up and run terrified from the room? On the contrary, the manly Mulder stands and fights. Grabbing the nearest weapon, he whacks at the mutant monstrosity with a fly swatter, killing it on the spot. Um . . . did I forget to mention that this real-life monster is only an eighth of an inch long?

Knowing this last little tidbit, most scriptwriters, even friends, would have smiled lightly while pointing toward the door. To my delight, Chris's eyes lit up. Finally, after years of earnest suggestions, I had come up with something that had sparked his interest.

My acquaintance with this tiny mutant monster began during my first week of graduate school at Indiana University. Dr. Thomas Kaufman from the biology department showed a series of slides that featured the most bizarre creatures I had ever seen. They were fruit flies with wonderful names like *Antennapedia*, the fly with legs where antenna should be, and *Proboscipedia*, the fly with legs coming out of its mouth. As a science fiction fan (then and now) I remember thinking that studying genetics was going to be very interesting.

After telling Chris Carter about the bizarre flies, he paid a visit to Indiana University and spent the day with Tom Kaufman and his laboratory of devoted *X-Files* fans (Kaufman included). Chris returned to Los Angeles with a film loop of a developing fly embryo, a scanning electron microscope picture of *Proboscipedia*, a diagram drawn by Kaufman explaining his science, and a T-shirt from a recent scientific meeting. Using his newly acquired knowledge and souvenirs, Chris wrote one of my favorite *X-Files* episodes: "The Post-Modern Prometheus." In a takeoff on the early Frankenstein movies, Chris invented a "mad scientist" who tinkered with the genes of fruit flies as a springboard to playing with the genomes of people. Chris implied in the episode that the fruit fly *Proboscipedia* was the product of genetic engineering. However, the real origin of *Proboscipedia* didn't involve the intervention of man. This weird mutant owed its deformity to random mutations. Because of a single defective gene, flies were born with legs coming out of their mouths.

Welcome to the wonderful world of genetics. A world where thousands of mutations can have absolutely no effect on an organism while one mutation in the wrong place can produce a monster. Mutations are random, accidental changes to the order of nucleotides in the DNA of a cell. Recall that there are four different kinds of nucleotide beads on a DNA chain. In regions of the DNA that are genes, the order of the four different nucleotides determines the order of amino acids in a protein. Change the succession of nucleotides in a gene and you can change the appearance and function of the protein that it specifies. Mutations can be as simple as a single nucleotide changing to one of the other three or as substantial as the loss or duplication of a large section of DNA. If an important protein gets altered because of a mutation in the DNA, flies can be born with legs dangling from their foreheads.

Can random, individual mutations cause legs to suddenly sprout from your head or turn you into a teenage mutant ninja turtle? Hardly. If a mutation occurs in one of your skin cells, maybe that cell will turn turtle green, but who cares? There are millions of normal skin cells nearby. The only mutations you have to worry about for your own health are ones that disrupt the ability of a cell to remain a good citizen of the body. If mutations transform a liver cell into a cancer cell, then that cell becomes a rogue criminal that no longer cares about the surrounding cells. The mutant cell begins to divide nonstop, producing other cells in its image that also incessantly divide. The gang of mutant cells crowds out normal cells and eventually becomes a tumor.

Mutations can sneak into DNA every time a cell splits in two because the cell must first duplicate its DNA. Once duplicated, the DNA is divided up so that both daughter cells receive the same amount of DNA as was originally found in the parent cell. The process of duplicating DNA is nearly, but not quite, perfect. Polymerases, the enzymes that copy DNA, have the Herculean task of making a perfect replica of DNA chains that contain millions of nucleotide beads. Every time an overworked polymerase makes a mistake and messes up the order of nucleotides without correcting its mistake, a new mutation appears in the DNA of that cell. The imperfection then becomes a permanent part of the cell's genome.

Random mutations in your egg or sperm cells won't affect you but might affect your children. This is because all 100 trillion cells of a child that develop from the mutant egg or sperm will have the mutation—a very different situation from a single cell in your body acquiring the same mutation. Even when they're found in every cell, however, most mutations are still inconsequential. To understand why, imagine that you are duplicating a mostly black drawing on a pretty good copy machine. The duplicated image is nearly, but not quite, identical to the original. Any imperfections, even large ones, which occur in black regions of the drawing won't be noticed. Only tiny specks that stand out against the white background will be visible.

Mutations in DNA are like the imperfections in the copy of a drawing. How consequential a new mutation is depends on its location in the DNA of the genome. Mutations in the 95 percent of the genome that is unused, "junk" DNA (the "black" region of the drawing) will scarcely be noticed. Mutations in the 5 percent of the genome that contains genes have a chance of affecting the organism. Yet even if a mutation in a gene is severe enough to produce a substantially altered protein, the effect on the cell can still be minimal if that particular protein isn't very important or if another protein can take its place.

And then there are the very important proteins—proteins that are the generals of the cell and without which cells can't function normally. When egg or sperm cells acquire mutations in genes that specify important proteins, severe birth defects can result. Some of these important genes are responsible for letting the cells of a developing organism keep track of which parts of the body are which. In the fruit fly, a gene called *Antennapedia* causes the cells near the top of the head to proclaim to other cells, "We're the top of the head!" The chemical message is picked up by proteins on the surfaces of nearby cells, causing antenna to grow in the proper place. Mess up the *Antennapedia* gene, and the cells near the top of the head incorrectly cry out, "We're the thorax, stupid, make some legs!" So legs appear where antenna should be. Humans also have the same important genes that lay out the segments of the body. Unlike flies, which can live quite happily with legs flopping from their heads, such mutations in humans only lead to live births on *The X-Files*.

If your only knowledge of mutations and mutants comes from comic books or television, you might think that most mutations are beneficial to an organism. Mutations in the media turn dumb turtles into wisecracking, crime-fighting, pizza-loving heroes, or give amazing, superhuman mental and physical powers to ordinary people who like to wear colorful skintight clothing. In truth, if you consider only mutations that have some effect on an organism, the vast majority of mutations are detrimental. Mutations tend to be harmful because it's easier to mess up a protein than improve it. If you put a gaggle of five-year-olds together with hammers and new television sets in a room and tell each child to give a TV a whack, chances are that reception will not improve. In the same way, just about every mutation will have either no effect on a protein or will make it worse. Protein function is rarely improved by random mutations.

This is the seeming paradox of evolution. Mutations are required to drive the evolution of an organism yet nearly all mutations are inconsequential or detrimental. Evolution, though, is not a paradox if you think of it as a force that is acting on a large population over a very long time. Somewhere, sometime, out of the millions of hammer-wielding five-year-olds, one will whack a TV and make it work better. Engineers will take it apart (the TV, not the kid), figure out what went right, and TVs all over the world will be produced in its image. Evolution has had millions of years to respond to the occasional beneficial mutation. A mutation that results in a better immune system would keep an individual alive in a world beset by pathogens. That person would be in a better position to pass on her genes to the next generation as others less fortunate perish from disease.

Even today, examples can be found of rare beneficial mutant genes that are driving evolution. As HIV continues to cause AIDS and death for large numbers of people worldwide, some individuals who are at high risk never become infected with the virus. Is it good fortune, or is there something in their genome? Scientists looking for connections among these lucky people found one. A mutant gene that helps protect people from HIV is found in the genomes of 10 percent of people descended from Northern Europeans. If two copies of the defective gene are present, as they are in 1 percent of this population, then these fortunate people are able to resist becoming infected with HIV. The mutation is in the gene for a protein that normally hangs out on the surface of immune cells and provides a doorway for the invasion of the virus. If this protein is damaged or not present, the virus has a very difficult time entering and infecting cells. This faulty but beneficial gene is not found in South Americans, Africans, or Asians.

So why, you might ask, are Northern Europeans so fortunate? Good fortune today may have come at the expense of a terrible catastrophe many years ago. For a mutant gene to become widespread in a limited population, something must have occurred where a person's very survival was dependent on having the mutation. Since infection by HIV was unheard of twenty years ago, that "something" was likely a different epidemic. With a reasonable estimate of the frequency with which one population produces another, and a pretty powerful computer, some scientists have calculated the date when the mutant gene was probably first present in more than just a few people. That date is approximately seven hundred years ago, the time when the bubonic plague was sweeping through Northern Europe, decimating the population. Bubonic plague, like HIV, is a disease of the immune system. Survivors of the plague either lived in isolated villages or had a mutant gene that conferred immunity to the plague bacterium. If the mutant gene was present in 1 percent of the population before the epidemic, it may have been found in 90 percent of the vastly reduced population after the plague killed off the people who did not have the gene. Some scientists believe that this same gene protects the descendants of plague survivors from HIV.

Another example of a beneficial mutation is a defect in the gene for hemoglobin. The hemoglobin protein is responsible for carrying oxygen from each breath to the cells of the body. Everyone has two copies of the hemoglobin gene. If only one copy is faulty, there are few consequences to the individual—the remaining good copy specifies enough protein to transport sufficient amounts of oxygen. If both copies are faulty, the person has a potentially life-threatening disease called sickle-cell anemia. About 10 percent of adults of African heritage have one good and one faulty version of the hemoglobin gene.

The benefit derived from having one defective version of the hemoglobin gene is now known. It protects against malaria. Malaria is a terrible disease caused by *Plasmodium,* a single-cell parasite that is transmitted to people by mosquitoes. Three hundred million people a year, mostly in Africa, are afflicted with malaria, leading to 2.7 million annual deaths. If you have one good and one bad version of the hemoglobin gene, you are much less likely to get malaria. It is not hard to imagine a time before modern medicine and mosquito control programs when an even larger percentage of the African population died each year from malaria. Those fortunate enough to have one bad version of the hemoglobin gene didn't get malaria and so lived longer than less fortunate people who had only "normal" hemoglobin. In the past, the benefits to an individual who had the mutant gene far outweighed the low chances that a child would inherit two bad versions of the gene and die from sickle-cell anemia. This is hardly a comfort to many African Americans who no longer derive any benefit from being protected against malaria yet have to experience the heartbreak of children with sickle-cell anemia.

Epidemics such as malaria, bubonic plague, and HIV are cited by some scientists as evidence that evolution doesn't proceed at a slow and steady pace toward a more "fit" organism. Rather, they believe that evolution progresses in running leaps due to catastrophic, unforeseen events that eliminate a large fraction of a population. Only those fortunate individuals whose genomes confer protection pass on their genes to future generations. This theory of evolution was first proposed by Niles Eldredge and Stephen J. Gould and is known as punctuated equilibrium. Devotees of *The X-Files* may recall Mulder referring to this theory in the episode "Leonard Betts."

In the realm of science fiction, mutations frequently lead to the generation of monsters. While most science fiction monsters tend to be fearsome mutant animals with a curious need to destroy New York or Tokyo, the monsters in *The X-Files* are mainly human. There is the Peacock family, whose monstrous exteriors hide a dark family secret; Victor, a translator of ancient Italian who brings new meaning to the term "high fat diet"; Leonard Betts, the mild-mannered emergency medical technician who doesn't mind losing his head if the situation demands it; and a janitor who is every woman's dream date. These human monsters provoke fear in viewers, but also sympathy. They did not choose their violent natures, they were born that way. The peculiar appetites and behaviors of *X-Files'* mutants are the unhappy consequences of defective genes. Although *X-Files* monsters kill, they take no pleasure in it. Killing is a means to survive—to eat, or to hide their mutancy from others who would destroy them.

When Mulder asks Dr. Polidori, the fictional Dr. Frankenstein–like scientist in "The Post-Modern Prometheus," why he created flies with legs coming out of their mouths, Polidori replies, "Because I can." After Chris's visit to Indiana University, he phoned to ask why Dr. Kaufman was studying those flies. I responded that understanding why flies become severely handicapped due to a single defective gene can help scientists learn how some genes control the fates of many different genes. By studying these types of mutants, scientists are uncovering the secrets of genetics and development. In "The Post-Modern Prometheus," Mulder asks Scully why Dr. Polidori is doing these experiments. Scully replies, "To unlock the mysteries of genetics—to understand how it is that even though we share the same genes, we develop arms instead of wings—we become humans instead of flies or monsters." I couldn't have said it better myself.

Genes 'R Us

In nearly every town, there is a place that can only be described as eerie. Perhaps it is the location of a dilapidated structure where no one ever seems to go in or out, or a yard where only the bravest of children will venture to retrieve an errant projectile. The Peacock property in the episode "Home" is such a place. When the audience first peers inside, a tortured woman is giving birth under barbaric conditions. As lightning flashes in the night sky, a horribly disfigured baby is born. The three shambling men that witness the birth pick up the crying newborn and bury it alive in a nearby field. Children playing baseball accidentally unearth the dead infant, whose appearance is so inhuman that the sheriff summons Mulder and Scully to the scene. Scully takes one look at the baby and is aghast to find that it suffered from multiple rare birth defects including Neu-Laxova syndrome, Meckel-Gruber syndrome, and extrophy of cloaca.

Scully's knowledge of medical genetics is vast indeed to have recognized the symptoms of these very real disorders since there are so few reported cases of Neu-Laxova syndrome or Meckel-Gruber syndrome. Perhaps Scully had recently studied the book *Smith's Recognizable Patterns of Human Malformation,* which summarizes a large number of human deformities. In an astonishing coincidence, two of Baby Peacock's rare abnormalities are featured back-to-back on pages 152 and 153 in this well-known compilation of human afflictions.

Neu-Laxova syndrome is extremely rare and always fatal. Scully must have noticed the small head with a sloped forehead, small protruding eyes, gaping mouth, and lack of chin or any hair. If the baby hadn't suffocated from being buried alive, it would have lived only briefly because of the massive central nervous system damage and incomplete brain. Meckel-Gruber syndrome is slightly more common, but just as fatal. Scully probably based her diagnosis on seeing the brain protruding from the skull, the short neck and extra digits associated with this heartbreaking syndrome.

Neu-Laxova syndrome and Meckel-Gruber syndrome are two of more than four thousand genetic diseases identified to date. For a disorder to have a genetic basis, the DNA of one or more genes must be different from normal. Genetic diseases are rarely the products of new mutations. Rather, they are the consequences of mutations in genes that occurred long ago—legacies from ancestors long dead.

More specifically, genetic diseases and immunities, and much else, are gifts from your parents. People inherit 3 billion nucleotides from Mom and another 3 billion from Dad. Join the 6 billion nucleotides of the human genome from end to end and the chain would reach over three feet . . . and it all has to fit into the nucleus of a human cell two hundred times smaller than the period at the end of this sentence. It helps that the human genome is not one massive chain of DNA but is divided into 46 pieces called chromosomes.

Chromosomes come in large and small sizes, depending on the length of their DNA chain. If you look into the nucleus of one of your cells, the DNA would appear long and stringy, much like a plate of spaghetti. Distinguishing individual chromosomes among the forty-six residents of the nuclear soup is like trying to trace a single spaghetti strand in a heaping helping, especially when covered with tangy cellular meat sauce. Analyzing chromosomes is easier when a cell gets ready to divide. At that time, each chromosome's DNA chain vastly condenses in size, making a chromosome look more like an individual sausage. Chromosomes at their most condensed can be easily seen with the help of a light microscope.

Getting a two-inch-long chain of DNA to fit into the tiny dimensions of a supercondensed chromosome is like trying to stuff a rubber band the length of a football field into the football. If you and a partner hold a 100-yard-long rubber band at either end and begin twisting it, the rubber band starts shrinking as it coils into a neat spiral pattern. As you keep twisting, the coils twist on top of previous coils, and the rubber band shrinks further and further in size. Finally, the coils are so compact that the rubber band condenses into a small object, about the size of a football. The compacting of DNA into a chromosome is a bit more organized, but it is essentially the same idea.

Scully sends away some of Baby Peacock's tissue to the FBI lab so that the infant's chromosomes can be examined. When Scully holds up the completed chart of the baby's chromosomes, you can see the supercompacted chromosomes from a cell that was moments away from splitting into two cells. Chromosomes in this compacted form look like Siamese twin sausages, connected somewhere along the length. The sausages are "twins" because the DNA has been duplicated, which occurs right before a cell is ready to divide. The DNA chains of all forty-six chromosome sausages must be duplicated so that a copy of each chromosome ends up in each of the two daughter cells. As the cell divides, the Siamese twin sausages break apart with one sausage ending up in one cell and the other ending up in the sister cell. This division of Siamese twin sausages into the two daughter cells occurs for all forty-six chromosomes each time a cell divides. The result is that every new cell has precisely the same number of chromosomes as the parent cell.

Chromosomes come in two varieties, known as autosomes and sex chromosomes. You have twenty-two different autosomes, numbered from 1 to 22 according to their size. Autosomes always come in pairs, so in every cell there are two chromosome 1's, two chromosome 2's, and so on. Along with your twenty-two pairs of autosomal chromosomes, you have two sex chromosomes. If you are a woman, you have two X chromosomes; if you are a man, you have one X and one Y chromosome. Twenty-three of your forty-six chromosomes were inherited from your mother and the remaining twenty-three came from your father. Each of your parents provided you with one of every autosome and one of your two sex chromosomes.

If you open up one of your almost-ready-to-divide cells and scatter your chromosomes on a piece of glass, you would see forty-six chromosomes of various sizes if you use a light microscope. The two largest chromosomes, which look like Siamese twin sausages connected in mid-length, are easy to spot. Those would be your two chromosome 1's. At the other end of the spectrum, you would see two tiny chromosomes connected near one end—those are your chromosome 22's. The sizes of your other chromosomes are somewhere between these two extremes. Some chromosomes are so similar in size and appearance that it is impossible to tell them apart without help. Determining which of your scattered chromosomes are number 9's and which are number 10's, for example, requires special treatment of the DNA so that a particular chromosome will fluoresce with a unique color.

The chart displaying Baby Peacock's chromosomes in the episode is typical of a chromosome chart. To produce this type of chart, a digital image is taken of color-treated chromosomes from a single busted cell by a camera hooked up to a microscope. With the help of a computer (which replaces the scissors and adhesive tape I used when playing with Polaroid pictures of chromosomes in college), chromosomes are "cut" and "pasted" for placement in pairs from 1 to 22. The two sex chromosomes are always last on the chart, either two X chromosomes together or one X paired up with the much smaller Y chromosome. The most frequent need to open up a cell and lay out the chromosomes on a chart is when couples want to know the sex of their baby before birth. If the baby's two sex chromosomes are X's, it is a girl; X and Y, and it's a boy. Two Y chromosomes, and it's an alien.

Scully wants to examine a chart of Baby Peacock's chromosomes since severe deformities in babies can sometimes be traced to visible changes in chromosomes. Many genetic diseases occur because a chromosome has lost a large chunk of DNA from one or both ends of the chain—DNA that contained important genes. Recall that a gene is simply a short region along the DNA chain that gets copied into the similar molecule, RNA, which is then used as an instruction sheet for making a protein. All proteins have a job to do, with some jobs being more critical than others. If a chromosome is missing a piece of DNA so large that the chromosome is visibly shorter than normal, then a number of genes will also be missing. When Scully analyzes Baby Peacock's chromosomes, she mentions that there is noticeable chromosomal breakage, meaning that whole segments of DNA chains have broken off and are missing.

Scully refers to Neu-Laxova syndrome and Meckel-Gruber syndrome as "autosomal dominant disorders." Scully doesn't explain this statement since it would have taken her the remainder of the episode. While "Home" is one of my favorite *X-Files*, some viewers may have preferred a lengthy soliloquy on genetics to learning the dark and disturbing secret of the Peacock clan.

Scully is correct when she tells Mulder that Neu-Laxova and Meckel-Gruber syndromes are autosomal diseases. This means that the defective genes resulting in these syndromes are located somewhere on autosomes and not on sex chromosomes. Scully doesn't say which autosomes contain the faulty genes because that information isn't known. Simply knowing that the disease genes are autosomal is like being told that San Antonio and Los Angeles are cities in the United States and not Mexico, only you aren't told which states the cities are in or how close the cities are to each other. Just a few thousand of the 100,000 human genes have had their precise positions on the chromosome "map" determined so far. Once the Human Genome Project is completed, scientists will be able to determine the precise locations of defective genes that cause rare genetic diseases like Neu-Laxova and Meckel-Gruber syndromes. It's important to keep in mind, though, that just knowing where a disease-causing gene happens to be on a chromosome doesn't mean that a cure is in sight. Knowing that Los Angeles is in California tells you nothing about the population, the weather, the schools, or whether it's a nice place to vacation. It only tells you where it's located.

When Scully explains to Mulder that Neu-Laxova and Meckel-Gruber syndromes are "dominant" genetic disorders, she is referring to the relationship between different versions of a gene. Since everyone has two of each autosomal chromosome, everyone has two versions of every gene found on an autosome. The two versions of any gene can be either identical to each other or different from each other. Let's consider the gene that determines the shape of your earlobes. Some people, like me, have loose earlobes. Other people, like my husband, have attached earlobes and consequently can't wear quite as many earrings. The gene that determines earlobe shape is located on an autosome so everyone has two versions of the earlobe gene. These versions can specify attached earlobes or loose earlobes. If you have two attached versions of the earlobe gene, you have attached earlobes; if you have two loose versions, you have loose earlobes. But what if you have one loose version and one attached version? For earlobe shape, the loose version of the gene is the more important one. It dominates the attached version. Versions of genes that dominate other versions are called *dominant*. The meek versions of genes are called *recessive*. This means that if you have one loose version and one attached version of the earlobe gene, you will have loose earlobes. Look in the mirror. If you have attached earlobes, you must have two attached versions of the earlobe gene.

This hierarchy of gene versions is common. Take hair color. If you have a brown version and a blond version, you will have brown hair. Brown is dominant to blond. Anyone who has blond hair either has two blond versions of the gene or makes regular visits to a hairdresser. Mr. Spock of the starship *Enterprise,* half human and half Vulcan, has pointed ears. This means that the pointed-ear version of the gene he inherited from his Vulcan father is dominant to the round-ear version of the gene he inherited from his human mother. When Scully calls Neu-Laxova and Meckel-Gruber syndromes "dominant disorders," she means that the defective version of the Neu-Laxova syndrome gene dominates the normal version of the gene. Having one defective version of any dominant disorder is enough to give someone the disease. It doesn't matter if you also have a normal version of the gene.

This is the story behind Huntington's disease, an autosomal dominant genetic disease that devastates the lives of about one in every ten thousand people. Everyone afflicted with Huntington's disease has a normal version of the gene on one of their chromosome 4's and a defective version on the other chromosome 4 that specifies a faulty protein that clumps when it shouldn't. Since parents contribute one of each of their autosomes to their children, a mother with Huntington's can contribute either her normal chromosome 4 or her disease-bearing chromosome 4 to her child. This means that there is a fifty-fifty chance of passing on the chromosome containing the defective gene to each child. The degeneration of neurons associated with Huntington's doesn't begin until a person reaches her thirties or forties, so whether or not one has inherited the faulty gene from an afflicted parent usually isn't known before having one's own children and possibly passing the defective gene on to them. Genetic testing is now available for people with a sick parent, but most choose to remain ignorant of their genetic fate.

If Scully is correct, and Neu-Laxova and Meckel-Gruber syndromes are dominant autosomal disorders like Huntington's, then Baby Peacock's father or mother must have had the defective genes. If the father or mother has the same lethal syndromes as Baby Peacock, and these syndromes are dominant, he or she would have been, ah, dead at birth. Is this another X-File?

Actually, the answer is more mundane. Scully, no doubt dismayed at the sight of poor Baby Peacock, makes a slight error when talking about the syndromes. They are in reality not autosomal *dominant* disorders but rather autosomal *recessive* disorders. To be afflicted with an autosomal recessive disorder, you need both versions of a particular gene to be faulty. If you inherit only one faulty version of a gene, you don't have the disease. Your normal version of the gene dominates the defective version. You are, however, considered to be a carrier of the disease. This means that you can pass on the defective version of the gene to your offspring. If you are fortunate and marry someone who doesn't have a defective version of the same gene, your children cannot inherit the disease. They can only become carriers like yourself. If you are unfortunate and marry someone who is also a carrier of the same disease, your children have a one in four chance of inheriting both your bad copy of the gene and your spouse's.

This is the genetic roulette of cystic fibrosis and other autosomal recessive disorders. Cystic fibrosis is caused by a tiny flaw in a gene on chromosome 7. This gene specifies a protein that normally makes tunnels through plasma membranes, the coverings that surround all cells. These particular tunnels are for the exclusive use by the chemical chloride, so that chloride can vacate the cell when necessary. A flaw in the gene for the tunnel protein means that the tunnel is either not made or it's defective. This seemingly innocuous transportation problem of a single chemical leads to respiratory and digestive problems that take an enormous toll on the health of the afflicted individual.

If a room contains twenty-one Caucasians, chances are that one of them is a carrier of the cystic fibrosis gene and doesn't know it. After all, that person will never show any signs of the disease. In the past, it was beneficial to have one faulty version of the cystic fibrosis gene since it may have offered protection against typhoid fever—hardly a comfort in the United States today. As long as a carrier of the cystic fibrosis gene marries someone with two normal versions of the cystic fibrosis gene, there is no chance that any of their children will have the disease, since the children will always inherit at least one normal version of the gene. But mates are chosen for reasons other than invisible genes. About one in every two thousand children born in the United States has cystic fibrosis. For most parents, the devastating diagnosis of their child is the first inkling that they are both carriers of the faulty gene.

Baby Peacock's problems go far beyond the autosomal recessive diseases of Neu-Laxova and Meckel-Gruber syndromes. Observant viewers of Baby Peacock's chromosome chart probably noticed that there are far more than the normal 46 chromosomes. Baby Peacock has 92 chromosomes—four copies of each chromosome instead of the normal two! When it comes to chromosomes, more is definitely not better. Millions of years of evolution have led to precisely the correct number and types of genes to form a living, breathing, usually rational person. Extra copies of genes are not good. It's like throwing an extra valve into a precision sports car without retooling the rest of the engine. Instead of getting a more powerful car, the result is a defective lemon.

Having even one extra chromosome can lead to very severe birth defects, when it permits the fetus to survive at all. Down's syndrome, a duplication of tiny chromosome 21, occurs in about one out of every one thousand births. Instead of the normal two copies of chromosome 21, people with Down's syndrome have three copies. Only two other autosomal chromosomes, numbers 13 and 18, can be duplicated and still lead to live births. Unfortunately, the severe physical problems associated with having extra versions of the genes found on these two chromosomes mean that survival is rare beyond the first year.

While extra autosomal chromosomes are normally a death sentence, extra sex chromosomes are not. There is already an unequal number of X chromosomes between men and women. Women have twice as many X chromosomes as men, yet survive the experience quite nicely. On the contrary, having only a single X chromosome is not always in the best interest of men. By having only a single version of every gene on the X chromosome, men are more likely than women to have genetic diseases that involve faulty X chromosome genes. The best-known genetic disease connected with the X chromosome is hemophilia. For women to have the disease, both of their X chromosomes need to have the faulty version of the gene. Men, however, have only a single X chromosome, which they inherit from Mom. If that single X chromosome has the defective gene, the son will have hemophilia. The gene for color blindness is also located on the X chromosome, which is why so many men and so few women are color-blind.

Scientists used to puzzle over how finely tuned human cells could function regardless of whether they contain one or two X chromosomes. The answer came in the mid part of the twentieth century when it was discovered that a random X chromosome in every woman's cells curls up into a little ball and becomes dormant. Most of the genes on the inactive X chromosome no longer function so that just like men's cells, women's cells have only one active X chromosome. This natural process, called X-chromosome inactivation, is the reason why having extra copies of the X chromosome is not a death sentence. Women can live with several extra copies of the X chromosome, since all but one would be inactivated in each cell.

X-chromosome inactivation is the secret behind beautifully colored calico cats. Ever wonder why all calico cats are female? The cat gene for black and orange hair color is on the X chromosome. Since male cats have only one X chromosome, they can have either the black version of the gene, and be black, or the orange version of the gene, and be orange, but they can't be both colors at the same time. A female cat can have both a black version and an orange version of the color gene on her two X chromosomes. During kitty embryo development, patches of black and orange color appear because in every one of her cells, only one X chromosome actually functions. If the chromosome containing the black version is inactivated in one cell, then the cell will produce an orange color due to the orange version on the remaining active X chromosome. All the cells derived from that cell have the same X chromosome inactivated. You can tell from the sizes and locations of the color patches on the female cat which cells are the direct descendants of a cell that first inactivated one of its X chromosomes.

If Baby Peacock's only problem was an extra X chromosome, it would have few obvious physical manifestations besides being taller and thinner than average. Baby Peacock's genetic problems are severe, however, by virtue of having two entire extra sets of chromosomes. Scully is so shocked to see all the extra chromosomes on the chart that she first concludes that someone in the lab must have botched the test. Babies rarely develop with one extra set of chromosomes so it is a miracle that Baby Peacock survived long enough to be born with two extra sets. Realizing that the chromosome chart is most likely accurate, Scully and Mulder propose separate theories for how Baby Peacock ended up with so many extra chromosomes.

Scully's theory for the extra two sets of chromosomes is a "maldivision of the centromere." The more correct scientific term for what Scully is referring to is "chromosomal nondisjunction." I'm glad that I was able to clear that up. However, for those of you who don't dust around your genetics degree every week, here's what Scully is talking about. All of us begin as a single cell, the fusion of egg and sperm sex cells. To generate these sex cells, cells called germ cells must go through the type of cell division called meiosis. During meiosis, daughter cells are produced that contain precisely half of the chromosomes—one of each autosome and one of the two sex chromosomes—of the original germ cell. That way, when the twenty-three chromosomes in the egg cell and the twenty-three chromosomes in the sperm cell come together during fertilization, a never before and never again in the history of man concoction of forty-six chromosomes join forces to create a new and unique person . . . like you.

Two out of every one hundred pregnancies start with a fertilized egg that mistakenly has three sets of chromosomes instead of the normal two. Scully's theory is that Baby Peacock has two extra sets of chromosomes because of a problem that occurred when a germ cell was dividing into two daughter sex cells. In some embryos that have an entire extra set of chromosomes, one of the original germ cells didn't divide into two daughter cells like it was supposed to. Only one daughter sex cell was produced, with all forty-six chromosomes instead of the normal twenty-three. Fuse this sex cell (say, it's an egg) with a normal sperm cell and the fertilized egg will have three complete sets of chromosomes—forty-six from the egg and twenty-three from the sperm. If both egg and sperm have an extra set of chromosomes, the fertilized egg will have two extra sets, which is Scully's explanation for what happened to Baby Peacock.

Mulder has a different theory. He believes that more than one sperm fertilized the egg. Although eggs have developed mechanisms to prevent fertilization by multiple sperm, these mechanisms are not always successful. Two thirds of the time when an embryo has three sets of chromosomes, the extra set is due to fertilization by two sperm. Otherwise, extra sets of chromosomes are due to mistakes in meiosis—Scully's theory. So the odds are that Mulder is correct.

Baby Peacock is the poster child for genetic disorders. The probability of so many genetic problems appearing in one child is virtually zero unless the family ignored religious and state laws and practiced inbreeding—the mating of couples who are closely related to each other. There are good reasons why laws exist against inbreeding. If your genome is similar to the average, you are a carrier of five to ten lethal recessive versions of different genes and a host of versions of other defective recessive genes that are incapacitating but not lethal. This isn't something to worry about for your own health since the normal version of each gene dominates over the disease version. The chances are also small that you will marry someone who also has one of the same defective gene versions, which is the only way that your children could inherit two versions of a disease gene and have the disease. This is not the case for a population where inbreeding is common. The more related people are to each other, the more similar is their genetic makeup, and the more likely that two bad versions of the same gene end up in the children.

Inbred human populations, usually involving marriage between cousins, are not an uncommon phenomenon. On nineteenth-century Martha's Vineyard, an island off the coast of Massachusetts, the town of Chilmark was a half day's journey from any other village. As a result, residents of Chilmark intermarried, which brought out a genetic type of deafness. More than one quarter of Chilmark's residents were deaf and everyone in town could converse in sign language. Chilmark is no longer isolated so incidences of deafness are now rare. Inbreeding can also occur when groups become culturally isolated, marrying only members of the same small religious, political, or ethnic group. This has happened with the Amish and Mennonite communities of North America, who suffer from a rare form of short-limbed dwarfism, and the Choctaw Indians of Oklahoma, who have a high prevalence of systemic sclerosis, a disease of the skin and internal organs.

The sheriff tells Mulder and Scully that the Peacock family have a long tradition of inbreeding. The Peacock brothers, however, were in a quandary since the last related woman, their mother, was thought to have died in a car accident ten years back. This information leads Mulder and Scully to suggest that the boys kidnapped a traveler to be the mother of the next generation of Peacock babies. Such a scenario would be unlikely, though, since some of Baby Peacock's disorders were recessive. The random abductee also would have to be a carrier of the same very rare disorders as her captors. The dark truth is revealed soon afterward. The mother of the Peacock boys survived the car accident. The older brother is both father and brother of the younger Peacock boys, and at least one of the brothers was the father of Baby Peacock.

Scully's original diagnosis of Baby Peacock's problems included the uncommon disorder "extrophy of cloaca." Extrophy of cloaca actually isn't a genetic disorder, but a nonfatal mistake in the early development of the fetus that leads to a series of lower abdominal deformities. Many children are born with birth defects unrelated to defective genes. For example, children of mothers that smoke during pregnancy can have low birth weights, respiratory problems, and mental retardation, none of which are caused by faults in the child's DNA.

"You are what you eat" is a common axiom, but while you were in the womb, you were what your mom was eating. The most notorious example of a chemical eaten by pregnant mothers that led to deformed babies was the drug thalidomide. Thalidomide was produced by the Chemie Grunenthal pharmaceutical company in West Germany in 1957. In forty-six countries, thalidomide was marketed as a nontoxic wonder drug—no side effects, no morning sickness, and a safe night's sleep for pregnant women. Insomnia turned out to be only one of the side effects of taking the drug. Chemie Grunenthal had tested thalidomide on animals, but in an astonishing omission by today's standards, the company never bothered to test the drug on pregnant animals. The extent of the thalidomide disaster was soon apparent. As many as twelve thousand babies were born without arms or legs or both. Scientists estimate that for every child who survived to birth, three more died in the womb.

There are probably thousands of baby boomers in the United States who owe their physical health or even their lives to Dr. Frances Kelsey, a scientist who works for the United States Food and Drug Administration (FDA). In one of the great early triumphs of the FDA, thalidomide was never approved for use in America. Dr. Kelsey was given thalidomide as her first drug approval decision because it should have been a simple vote of yes. After all, the drug was being used in Canada and all over the world. Dr. Kelsey aggressively fought a one-woman crusade against the approval of thalidomide. The pharmaceutical company could not answer her questions on how the drug worked, and without that information, Dr. Kelsey thought that thalidomide was potentially too dangerous for distribution. It wasn't until 1961 that a link was made between as little as a single dose of thalidomide and birth defects. When tests of the drug were finally conducted on pregnant animals, the same types of rare birth defects were found. It is now known that thalidomide works by stopping the development of blood vessels, which guide the growth of limbs and organs. Thalidomide is still used in some countries to successfully treat leprosy, certain types of cancer, and AIDS. The FDA has recently approved the use of thalidomide in the United States for the treatment of leprosy, but only under highly restrictive conditions.

It is not clear why the Peacock baby suffered from nongenetic as well as genetic disorders. Lack of prenatal care, poor nutrition, and the unsanitary conditions in the house may all have played a role in the nongenetic birth defects. The genetic disorders were clearly the consequences of years of inbreeding and the inheritance of multiple lethal versions of genes. The laws of genetics were also not kind to the three Peacock boys. One brother has no scalp hair, another has patchy hair. All seem to have some overgrowth of facial bones, and walk with stooped posture and shambling gaits. None of them speak and they don't seem to feel any pain. After a final showdown with Mulder and Scully, only one brother and the mother remain alive. Before Mulder and Scully can stop them, mother and son drive off looking for a new home where they can continue passing on the Peacock genes.

Like Father, Like Son

According to the supermarket tabloids, aliens have been living among us for quite some time. Walk past most checkout counters and headlines such as *SPACE ALIEN ATE MY CHILD, ADMITS PROCTOLOGIST*, *UFO FOUND ON JUPITER!—SHOCKING TAPES REVEALED*, and *ALIENS FROM PLUTO LANDED IN MY LIVING ROOM AND TORTURED THIRTY-NINE BANANAS* vie for your attention. Only people who see aliens behind every unusual phenomenon could possibly take these stories seriously.

In the *X-Files* episode "Small Potatoes," Mulder drags a reluctant Scully to West Virginia to investigate the tabloid headline *Monkey Babies Invade Small Town.* The story describes the births of five babies with tails in a five-month period. Scully tells Mulder that a fetus normally develops with a tail but it is usually lost early during fetal development. Five children born with tails in a small town is highly unusual. Although Scully doesn't mention the precise numbers, only about twenty-four babies have been born with tails since 1884. In seven of these cases, the babies could actually wag the tails when crying or coughing, just like the babies in this episode. Since the persistence of tails in babies doesn't seem to have a genetic basis, Scully thinks that the defect in the West Virginia babies is due to a nongenetic disorder. Investigation is warranted, but by the Health Department, not the FBI. Scully suspects that Mulder has more interest in the second part of the headline, *Did West Virginia women mate with visitors from space?*

At the hospital, Mulder and Scully interview the latest mother, who names Luke Skywalker as the father. The mother asks Scully if Luke could also be the father of the other children. Genetic tests of the DNA of her baby and the other four "monkey" babies indicate that one man did indeed father all the children, but not necessarily Luke Skywalker.

The genetic test used in "Small Potatoes" and other *X-Files* episodes to determine parenthood uses a laboratory method called polymerase chain reaction, or PCR. PCR is commonly used to test for parentage since it can determine if a child's DNA is a combination of the DNA of the parents. In criminal investigations, PCR can also be used to match two DNA samples to each other. PCR is simply a method for making copies of a segment of DNA. Millions of copies. Enough copies so that the DNA becomes visible after applying a stain. Imagine that there is a piece of thread on the floor of your room—you probably wouldn't see it. Now imagine a million copies of the thread together on the floor. You couldn't miss them. It's the same with DNA. Before you can analyze two DNA samples to see if there is a match, you need to be able to see them.

It's probably not obvious why being able to see DNA can tell you something about parentage. The key is the choice of DNA sections that are copied. Some regions of DNA tend to vary between people, and these are the regions chosen to be copied in a PCR test. Many such variances in people's DNA are due to different numbers of repetitions of particular nucleotides in "junk" regions of the genome between genes. Consider, for example, the following sentence: THE DOG AND THE THE CAT ATE THE BAT. Each of the letters is like a nucleotide in a DNA chain. There are nine letters between "DOG" and "CAT," since the nucleotide letters T- H- E are repeated twice. Now take another similar sentence: THE DOG AND THE THE THE THE CAT ATE THE BAT. In this sentence, there are 15 letters between "DOG" and "CAT." If you were to make a copy of the segment of the sentence between DOG and CAT, one would obviously be longer than the other. The more repeats of the letters T-H-E, the larger the segment produced. The same is true for segments of DNA.

Scientists have identified a number of regions in the human genome that are different sizes in unrelated people because of different numbers of repetitions of three nucleotides. If you examine the sizes of DNA pieces from several of these regions, a unique profile can be generated for every person. If the PCR test is run on a baby and mother's DNA, the mother's contribution to the baby's DNA profile can be factored out, leaving only the contribution of the unknown father. For the five children born with tails, Scully learns that they all shared an abnormally small PCR fragment on chromosome 8 that the mother did not have. It could only have come from the mysterious father.

Mulder and Scully visit a clinic where four of the five mothers received insemination therapy. When a janitor at the clinic bends over while working, Mulder notices a scar indicative of a removed tail. The janitor immediately replaces Luke Skywalker at the top of the list of father suspects. When it's discovered that the janitor's father also had a tail, a genetic cause for the birth defect becomes more clear. Even Mulder discards his "fathered by an alien" theory in favor of simple human parentage.

Although Scully does not expound on the genetics of the tail trait, there is enough information provided in the episode to determine that the tail is due to an autosomal dominant genetic defect. How do I know this? First, the tail defect must be on an autosomal chromosome and not a sex chromosome because it was passed from father to janitor son and from son to at least one baby daughter. This couldn't happen if the tail gene was on the X or Y sex chromosomes. Fathers pass down their Y chromosome to sons and their X chromosome to daughters. If the trait is on the Y chromosome, only sons could inherit the defect, and the latest baby wagging a tail is a girl. If the defective gene is on the X chromosome, then the father of the janitor could not have passed the chromosome to his son because sons only inherit their father's Y chromosome. The gene therefore has to be on an autosome to be passed down from father to son and from son to daughter (see how handy a Ph.D. in genetics can be?).

The dominant nature of the tail trait can be deduced from the inheritance of the tail in successive generations. If the trait is dominant, then only one bad version of the gene is required to have a tail. If the trait is recessive, then all the mothers must also have the very rare "tail" version of the gene, an unlikely scenario unless inbreeding was rampant in the town. A dominant trait means that the janitor has a fifty-fifty chance of passing down the "tail" gene to his children. Since five children were identified with tails, there were probably more unsuspected offspring of the janitor that did not inherit the tail.

The mothers of the children had no idea that their husbands

were not the true fathers because the janitor could change his appearance and masquerade as other people, including the intrepid Luke Skywalker. An autopsy of the janitor's deceased father indicates that he had an additional muscle that Mulder believes allowed him and his son to morph their appearance and vocal cords. (If you're waiting for a scientific explanation for morphing, you'll be waiting a long time.) Scully wonders if there is a link between the "morphing" gene and the gene that results in a tail. She is not suggesting that these are manifestations of the same defective gene. Rather, gene linkage refers to the location of genes on the chromosome map. If two genes are located near each other on the DNA chain in one chromosome, then they will be inherited together when that particular chromosome ends up in a sex cell. This means that the five babies in the hospital with tails also have inherited the ability to morph their appearance. Given the difficulties of raising a child whose appearance doesn't change, one can only pity what the parents of the small potato babies will be facing in the years to come.

Food, Glorious Food

INT. INTERROGATION ROOM—DAY

                    VIRGIL
You look at this hideous monster...but I was only feeding a
hunger.

                    SCULLY
You're more than a monster. You didn't just prey on their bodies—you preyed on their minds.

                    VIRGIL
My weakness was no greater than theirs. I gave them what
they wanted. They gave me what I needed.

                                              —"2Shy"

A man and a woman, corresponding by means of an impersonal keyboard and screen, decide to take their relationship to the next level and meet in person in the *X-Files* episode "2Shy." A glance at the man and you can understand why at he would want to correspond in a non-visual manner. His skin contains patches of dry scales that a woman interested only in outward appearances might find distasteful. The woman, Lauren, no runway model herself, looks past his exterior because of the beautiful words he communicated to her by modem and now in person. Together in a car, isolated from prying eyes, they meet, they kiss, and a beautiful new relationship blossoms. . . . Of course if that's what really happened, *The X-Files* would hardly have lasted into its seventh season. So back up. They meet, they kiss . . . and a thick, gooey liquid spews from his mouth into hers. Lauren struggles to break the kiss as the slobbery substance burns her insides like acid. The man continues to hold their lips together as he sucks out her jellified innards. Now, that's romance on *The X-Files*.

By the time Mulder and Scully arrive on the scene, Lauren's body is nearly completely dissolved. Only a thin layer of goo surrounds her brittle skeleton. Scully analyzes some of the material responsible for Lauren's meltdown and discovers that the slime is mostly organic with a high content of hydrochloric acid. She tells Mulder that the material is similar to that secreted by the gastric mucosa, only somewhat more acidic. Scully also finds traces of pepsin, which she describes as a digestive enzyme.

Scully makes a slight mistake when referring to the gastric mucosa as the producer of pepsin and hydrochloric acid. Cells of the gastric mucosa actually secrete mucus, which forms a protective barrier for the stomach. Protection is required because nearby cells do secrete hydrochloric acid and pepsin, which would be very harmful to an unprotected stomach.

Pepsin and hydrochloric acid are two important agents involved in the digestion of food proteins. Everything that enters the body through the mouth—starches, proteins, fats, Lauren—must first be broken down into simple chemical building blocks. Eventually, these rudimentary chemicals travel through the bloodstream and are taken up by cells for use as raw materials or fuel. The stomach remains busy for about one and a half days digesting the pizza that was wolfed down in five minutes the night before. Over the course of a year, half a ton of food must be methodically reduced by the stomach to simple chemicals.

Pepsin, the enzyme Scully cites as present in Lauren's dissolved tissue, is a type of enzyme called a protease. Proteases have recently achieved fame as targets of the latest drugs to combat HIV, the virus that causes AIDS. The job of a protease is to break the chemical bonds between amino acids in proteins. Imagine that a protein with its chain of amino acids linked one after another is like a string of different-colored beads. The number of beads on the string, like the number of amino acids in a protein, can vary from a few to hundreds. Let's say that special scissors can cut the beaded string only when a blue or red bead is to the left of a green or yellow bead. Another scissors can cut only when a white bead is to the left of a red or black bead. The two scissors are analogous to two different proteases. A protease can cleave within a protein only when it recognizes particular amino acids in a specific order.

The protease associated with HIV precisely clips a large virus protein into several smaller proteins, and only these smaller proteins can function in virus reproduction. The new drugs in the battle to conquer HIV, the protease inhibitors, stick to the viral protease and keep it from working. If the protease is disabled, the smaller viral proteins are never generated and the virus cannot reproduce itself. Pepsin, the protease made by your stomach cells, is quite different from the HIV protease, which chops only one specific protein. Pepsin needs to cut into little pieces the huge assortment of proteins that enter the mouth on a daily basis. Pepsin accomplishes its mission because it can snip between many different pairs of amino acids.

At first glance, it might seem intuitive that the stomach needs hydrochloric acid to help "dissolve" food. This is not, however, the main reason for stomach cells to make acid. Acid strong enough to digest proteins—strong enough to break the chemical bond between amino acids—would have your innards looking like the late Lauren's in no time. One reason to produce stomach acid is because pepsin can't work without it. Pepsin is initially made inside stomach cells in a form that is immature and no danger to any protein. This nonfunctional state keeps pepsin from immediately chopping into little bits the proteins in the cell that made it, which would promptly kill off such a stupid cell. To keep this from happening, pepsin becomes functional only after it encounters hydrochloric acid in the stomach, which happens only after pepsin has traveled outside the cells where it is made. Freed from the confines of cells, pepsin can concentrate on mincing ingested proteins.

Mulder is naturally curious as to why anyone would want to liquefy Lauren with digestive juices. He asks Scully if the melted woman is missing anything in particular that she possessed when alive. Scully replies that all of Lauren is present with one glaring exception. Lauren, who was listed at 165 pounds, now weighed in at a svelte 122 pounds. This sudden weight loss was achieved by the nearly complete elimination of all adipose—the tissue that contains fat—from her body. Furthermore, Scully tells Mulder that Lauren is also missing all of her oils (liquid fats) and fatty acids (the type of fats found in membranes).

While the elimination of all fat from the body seems like a worthwhile goal to many, chances of surviving in that condition would be about as likely as reanimating Lauren. Fats and oils are essential members of a class of molecules called lipids. Without fats, cells would have no membranes to surround them; there would be no insulating sheaths around nerve cells; there would be no steroid hormones like estrogen and testosterone. Even the dreaded cholesterol, made by animal cells from fats, is a required component of membranes and life. As important as fats and oils are to the body, it is a mystery to Mulder and Scully why anyone would want to melt someone for their body fat. People normally eat foods containing fats—usually more than they need—and can also make additional fats in their cells. What could anyone want with forty-three pounds of liquid human fat?

Mulder and Scully learn more about the killer by examining another victim, a previously overweight prostitute, now just melted skin and bones. The struggling woman had scratched skin off the killer's arm, which Mulder sends to the lab for analysis. While the DNA results do not reveal a match in the FBI database, the skin sample is nonetheless very informative. Mulder and Scully are astonished to discover that the killer's skin sample contains no oils or fatty acids. Mulder theorizes that the killer requires fats to survive since he doesn't make any himself. As Scully puts it, the killer is a vampire who sucks fat instead of blood.

Mulder believes that the killer secretes a digestive substance into the victim's mouth that dissolves fats but leaves other body constituents alone. Given the abundance of different fats in the body and their widespread uses in cells and tissues, it is interesting to speculate on what that digestive substance might be. Most cellular material, such as proteins, sugars, and nucleotides, dissolve in water and not organic solvents like chloroform. Pure fats and oils can dissolve in chloroform but not water. There is a problem, though. When dealing with fats in living tissue, chloroform alone will not dissolve them all. Many fats, like the fatty acids that make up membranes, are attached to other substances that don't dissolve well in chloroform. So what is a fat-sucking vampire to do? Possibly use a mixture of three parts chloroform to one part methanol, a common recipe used by scientists who need to separate fats from other tissue ingredients (you can imagine the looks I received from colleagues when I explained why I needed this information). A fat-sucking killer could spew acid and pepsin into the victim to help break up internal tissue, followed by a chaser of chloroform and methanol to separate the fats. Then it's sucking time.

Mulder asks Scully if this habit of predigesting food is a common theme in nature. Scully gives Mulder an excellent example: scorpions. Scorpions are real-life miniature monsters, eating insects and spiders and drinking their blood. Scorpions can be rather bad-tempered since their food source is so rare that they are often compelled to wait months, or even up to a year, between meals. They grab their prey with large pincers and tear it apart with their mouths. Then, just like the killer in "2Shy," scorpions spew enzymes and other salivary substances onto the prey, softening it prior to dining. If the prey doesn't happily submit to being torn apart and predigested, the scorpion can paralyze or kill it with a sting from its lethal tail.

There is a second example from nature that Scully could have given Mulder, that of the sea star, also known as a starfish. Anyone who has walked along a tide pool at low tide is well acquainted with these colorful, multiarmed creatures. Knowing how they snack on barnacles and clams may make you look twice at a starfish in the future. Starfish eat by extruding their stomachs through their mouths. The inside-out stomach is pressed against food and enzymes are secreted to tenderize the repast. The stomach and food are then swallowed and digestion continues. It is doubtful that "2Shy" could have passed muster with the television censors if the fat-eating killer had table manners closer to starfish than scorpions.

Since it is unlikely that Lauren would have gotten into a car with a giant scorpion or starfish, even on the sort of blind date that one has arranged via e-mail, Mulder proposes that the killer represents a genetically different creature derived from humans. In other words, a human mutant. Someone who is unable to manufacture his own fats, but has to rely on others for his needs. Since the major form of fat in the body are fatty acids, a mutation in the biochemical pathway that makes fatty acids from simpler chemicals would cause an individual to be unable to produce most fats. Most fatty acids are built in cells by a single metabolic pathway that requires seven different enzymes. The product of the pathway is the fatty acid called palmitate. Other fatty acids are derived from palmitate by the action of additional enzymes. If the killer had a mutation in both versions of a gene for any of the seven enzymes required to make palmitate, he would have few fats in his body and would need to ingest fats from others.

With that said, I wouldn't spend time worrying about that next kiss. Anyone who couldn't make their own fatty acids wouldn't develop from the initially fertilized egg. The first division of the first cell requires the production of membranes of which fatty acids are the major component. While Mom provides some fatty acids during development of the fetus, a whole battery of mutations would be needed for her to provide all the fats necessary for proper growth. Another slew of mutations would be required once the baby is born to produce and sequester the chloroform and methanol necessary to acquire fats from others. The need for so many mutations in so many genes to generate so many new functions will thankfully confine such mutant monsters to the safety of your television screen.

Mulder and Scully finally catch up with the killer, who is revealed to be Victor, a translator of ancient Italian with a flair for making lonely women feel needed. Victor admits to killing forty-seven additional women but is completely unrepentant. He tells the agents that while in their eyes he is a monster, he was just eating to survive. In a sense Victor felt like Typhoid Mary. The poor woman was just making a living; it wasn't her fault if people died from her cooking. Although Victor would likely get the death penalty for mass murder, even with prison cuisine he probably wouldn't last long enough to order his final meal.

How to Get a Head Without Even Trying

INT. PATHOLOGY LAB—DAY

SCULLY

Case number 2268-97, Leonard Betts. As the remains are incomplete, all observations refer to a decapitated head, weight
11 pounds, 2 ounces.

Scully makes a visual examination of the head, feeling it with her gloved fingers—we stay on her face for this. She frowns.

SCULLY

The remains show no signs of rigor mortis or fixed lividity. Nor
do the corneas appear clouded. This would seem inconsistent
with the witnessed time of death, now nineteen hours ago...

Scully isn't sure what to make of this. She reaches for:
A SCALPEL
Which flashes in the light as her gloved fingertips pick it up.

SCULLY (O.S.)

I'll begin with the intermastoid incision...

CLOSE ON SCULLY

As she places the tip of her scalpel behind the ear, preparing to draw it over
the top of the scalp. Suddenly—

—Betts' EYES OPEN. Startled, Scully drops her scalpel—it hits the floor with
a CLING!

—"Leonard Betts"

Leonard Betts, in the episode named for him, is a quiet, unas-
suming man who works as an emergency medical technician and has
a knack for diagnosing cancers. While on an emergency run, his ambulance crashes violently into a truck and Leonard's head is severed

from his body. While decapitation would pose a problem for most people, it is only a temporary inconvenience for Leonard, who soon disappears from the local morgue. As usual, Scully's view on the missing body—modern-day body snatchers—is slightly more plausible than Mulder's explanation: Leonard Betts's headless corpse kicked its way out of the morgue freezer cabinet and walked out by itself. When Scully examines Leonard's head, which was left behind, she is astonished to see it open its mouth and blink its eyes. Scully wonders if this frightening incident was caused by residual chemical activity stored in cells, or was there something very strange about the deceased Mr. Betts?

Scully and a pathologist find clues to just how strange Mr. Betts is when examining a slice through the frontal lobe of Leonard's brain: there is evidence of an enormous glioma. Gliomas are tumors of glial cells, which make up the supporting tissue of the brain. Since over half of all brain tumors are gliomas, simply finding one in Leonard's brain is not unusual. What mystifies Scully and the pathologist is that every cell in the brain tissue segment appears cancerous. The brain material probably looked highly abnormal to them since cancer cells no longer participate with neighboring cells to create nice, ordered tissues. Scully and the pathologist may also have seen cells with abnormally large nuclei (the nucleus contains the cell's DNA genome), due to an increased amount of DNA that is present in many cancer cells. Some of the cells were also probably in the process of dividing when the head was severed and thus had highly condensed and visible chromosomes. Such condensed chromosomes would never be seen in normal brain cells, since normal brain cells never divide. Cancer cells, on the other hand, divide frequently. If Scully and the pathologists see abnormalities throughout Leonard's brain, they would conclude that his brain was a mass of cancer cells.

While Scully is examining Leonard's head, Mulder pays a visit to Leonard's apartment, where he finds evidence of recent use. A mysterious black liquid is in the bathtub, which Mulder identifies from an empty bottle as providone iodine. Scully thinks that Mulder has finally lost his senses when he tells her that Leonard minus his head has been making himself at home. If Mulder had reached into the bathtub, he could have been more explicit in his explanation to Scully. Leonard was in the tub regenerating his missing head.

Mulder hypothesizes that Leonard Betts's cancerous condition might explain his regenerative powers. He suggests that cancer cells might be part of Leonard's normal state, which allows him to regenerate limbs, or even his entire body. Mulder backs up his contention by noting that providone iodine is often used by scientists to help amphibians regenerate. Scully scoffs at his explanation, saying that while salamanders can regenerate missing body parts, this is beyond the ability of any mammal. While Scully is accurately relating what was known at the time the episode was written, she probably wouldn't make such a sweeping statement today.

Since the mid-eighteenth century, scientists have been studying how amphibians such as newts and salamanders regenerate sliced-off tails, limbs, jaws, and even eye tissues. Less noticeable but equally important to the amphibians is the ability to regenerate heart muscles and spinal cords. The process that amphibians use to accomplish such feats of self-repair is becoming clear as scientists compare regeneration of missing limbs with the natural generation of the same limbs in an embryo. In the early stages of development from a fertilized egg, an organism's cells are immature; their fates can still be changed. This was discovered from experiments on frog embryos. When a frog embryo is only a few hundred cells old, a piece of tissue destined to become skin can be cut away and transplanted to the "brain" region of another embryo. The grafted tissue eventually becomes part of the brain. If a piece of future skin is removed from an older frog embryo, it still becomes skin even if transplanted into the "brain" region. The fate of the cells were already determined. For most organisms, the route toward cell specialization is a one way-street, with cells maintaining their specialized identities until they die.

Amphibian cells have learned the secret of turning back the clock, of losing their specialization and reverting back to a time when any fate was possible. When a newt or salamander loses a leg, the wound is healed and the process of growing back the limb begins. In a manner that is still poorly understood, bone, skin, and blood cells at the site of the wound regress back to a time when they had no identity. This newly generated mass of virgin unspecialized cells, called a blastema, then starts rapid cell division. Chemical signals are received by the growing mass of blastemal cells, causing them to learn new roles and form the missing limb.

Outside of Mr. Betts, only two types of human cells are known to regenerate—blood cells and liver cells. This regeneration is different from what occurs with amphibians. When a mammalian embryo is developing, a few cells are set aside before they become specialized. These cells, called stem cells, are able to replenish blood lost to wounds or liver cells that die during the normal course of life. Bone marrow also contains stem cells that, at least in plastic dishes, can become all sorts of different cell types—muscle, fat, bone, or cartilage—depending on what nutrients they are given. Very recently, Italian scientists found that if they injected bone marrow cells into the bloodstream of mice with damaged muscles, the cells traveled to the site of the damage and became new muscle cells. What is true for mice, though, may not be true for humans. It is well known that adult human muscle cells once destroyed are not replaced. Stem cells have also been found in the mammalian central nervous system. These cells, when combined with various nutrients, can change into any of the three major cell types of the adult brain, at least in the laboratory. What role the stem cells of the central nervous system play inside a mammal isn't known.

So what are the chances that regeneration of human body parts will leave the world of science fiction and enter the world of science? Not as remote today as they were when Scully told Mulder that mammals can't regenerate. This leap in our understanding of what the future may hold comes from a completely unexpected observation by a scientist who was studying the immune system. Ellen Heber-Katz of the Wistar Institute in Philadelphia made a routine request of her technician: poke holes in the ears of laboratory mice so that they can be tagged for identification at a later date. When Dr. Heber-Katz was ready to insert the tags a few weeks later, not a glimmer of the holes could be found. The wounds were not merely healed, it was as if the holes had never existed in the first place. Thinking like Scully, Dr. Heber-Katz made the logical assumption that the technician had simply forgotten to make the holes. A few weeks after new holes were made, they had again disappeared without a trace.

These strange happenings opened the mind of Dr. Heber-Katz to a remote but tantalizing possibility: What if the mice had regenerated the tissue and cartilage to fill in the holes? Looking closely at the process of how the mice ear tissue filled in the gaps, she noticed the presence of blastema, the same masses of unspecialized cells found in amphibians that are re-forming limbs. The mice seemed to be undergoing genuine regeneration, even though mice, like humans and other mammals, shouldn't have this capability. Dr. Heber-Katz next tested if any other parts of a mouse could regenerate. Clipping off a half inch of tail led to an amazing 75 percent regeneration.

You are probably expecting me to tell you what happened when she chopped off a leg. The reason why I don't have this information is somewhat obvious: without cauterization, which would invalidate the experiment, the mouse would die from massive blood loss before there was even a remote possibility of regeneration. So unfortunately, the full extent of the regeneration abilities of these mice is not known.

A disclaimer is probably needed at this point: DO NOT ATTEMPT THESE EXPERIMENTS ON YOUR PET MOUSE! While Dr. Heber-Katz's unusual healer mice are not visitors from the void, they are also not ordinary mice. The mice that regenerate had damaged immune systems and were being used to study diseases that involve the immune system, like multiple sclerosis. Amphibians that regenerate also have virtually no immune systems. A key experiment was performed on the healer mice that pointed to their malfunctioning immune systems as the reason why they were able to regenerate. Adult healer mice, which no longer regenerate as quickly or as well, were treated to destroy their T-cells, the immune cells from the thymus. Then the adult mice healed just fine.

Dr. Heber-Katz's speculations as to why mice have a cryptic ability to regenerate are fascinating. What if mammals like mice and humans share with amphibians the genes necessary to regenerate missing tissues but that T-cells in some way keep these genes from functioning? She believes that organisms originally had two ways of healing wounds—an immune system and regeneration. Sometime during evolution, when both systems were becoming more complex, the immune system and regeneration became incompatible with each other and a choice was necessary. While regeneration may at first glance seem like the better choice, T-cells are the body's main weapon against tumors. What would be the point of being able to replace a lost arm if the body succumbs at an early age to rampant malignant cancers? Dr. Heber-Katz believes that the very immune system that protects us from pathogens and cancers is suppressing our natural ability to engage in self-repair.

It's tantalizing to imagine a day when science can alter cells near a damaged spinal cord or amputated limb such that regeneration and a functional immune system are not at odds with each other. No one would fear amputation because the limb would simply grow back. Heart muscles damaged by disease could re-form like new, and spinal cord injuries would no longer confine victims to wheelchairs. Leonard Betts, having regenerated his head, suffers only temporary pain and inconvenience later in the episode when he tears off a thumb to free his hand from handcuffs. He knows that the thumb will grow back.

While regenerating limbs or a spinal cord may one day be science and not science fiction, regenerating a head, as Scully points out, is another issue entirely. Mulder correctly reminds her that worms can regenerate their heads. Chop a one-inch planaria worm into three hndred little pieces and in a few days, three hundred new worms are swimming around. Worm regeneration involves the same type of blastema as found in amphibians and the healer mice, but the blastema form in a different manner. Instead of cells near the wound reversing the clock and losing their identities, worms contain cells called neoblasts throughout their bodies that remain in an immature state. As long as a neoblast cell is present in a chopped worm segment, it can start dividing and generate the cells necessary to "fill in the gap," even if that gap includes the head and most of the worm. Scully is correct to remind Mulder that while worms can regenerate their heads, Leonard Betts is no worm.

So what is Leonard Betts? The possible relationship between the immune system and regeneration throws a whole new light on the mystery. Leonard's immune system must be highly compromised to allow so many cancer cells to be present, and a damaged immune system may be necessary for his regeneration abilities. It is also possible that Scully and the pathologist mistook blastema for tumors, and that Leonard's innate ability to regenerate comes from the ability of his cells to easily lose their identities, form blastema, and then gain new identities.

This possibility seems more likely than Mulder's suggestion that

Leonard is a mutant whose every cell is predisposed toward cancer. Why my theory is somewhat more plausible requires a brief digression into the nature of cancer. Cancer cells, like blastemal cells, have lost their specialized nature, but unlike blastemal cells, they are not capable of ever regaining an identity. Cells form blastema without any permanent changes to their DNA—in other words, no mutations are necessary. For a cell to become a cancer cell, mutations need to occur in many genes in that one cell. Once these genes are mutated, getting rid of the mutations and reverting back to normal isn't an option.

Healthy cells don't become cancer cells by acquiring mutations in just any genes. Cells are kept on an even keel, dividing when they are supposed to and cooperating to form tissues in the body, by two important sets of genes: proto-oncogenes (precancer genes) and tumor suppressor genes. Once mutations occur in some mixture of these genes, cells irrevocably change into cancer cells.

The function of a normal proto-oncogene is to help the cell understand when it is time to grow and divide and when it is time to rest and relax. Normal cells are highly disciplined. They divide only when they receive the proper signal that the time is right, which is relayed through a series of commands beginning with a protein on the surface of the cell's plasma membrane. The surface protein acts like a light switch. When in the normal "off" position, the cell is at rest with no thoughts of dividing. When the surface protein comes in contact with a specific substance—a growth factor—floating in the bloodstream, the protein is switched "on." Orders are then passed from this activated protein to a series of other proteins in the cytoplasm and finally to proteins in the nucleus. The nuclear proteins turn various genes on or off, which sets the cell on the path toward splitting in two. When the growth factor that started the process is no longer in the bloodstream, the surface proteins stop contacting the growth factor and revert to the "off" position. Signals cease being transmitted down the chain of proteins and the cell stops further divisions.

If any of the proteins in this chain of command decide to ignore orders from above and incessantly transmit a "go for it" signal to the remaining proteins in the chain of command, then the cell is instructed to divide and divide and keep dividing regardless of whether it is really supposed to be doing so. Through enormous strides in recent research, it is becoming clear that many of the genes for the command-chain proteins are proto-oncogenes. When a proto-oncogene is mutated, it is converted into a cancer-causing gene known as an oncogene. A defective protein is produced from the oncogene that permanently relays orders for the cell to divide. Normal cells have two undamaged versions of every proto-oncogene. It only takes one mutated version of a proto-oncogene to start a cell down the dangerous route of unchecked cell division, the hallmark of a cancer cell.

If the DNA in Leonard Betts's cancer cells was scrutinized, more than one proto-oncogene would probably be found in mutant form. Cells that become cancerous and form tumors usually have several mutated proto-oncogenes. Scientists interpret these results to mean that mutations are required in a number of different genes in a single cell to lead that cell down the path toward cancer. Cells with mutant proto-oncogenes are like cars with one foot permanently on the accelerator. Fortunately, there is still a second foot on the brakes. The brakes are applied by the protein products of tumor suppressor genes. These genes are aptly named because the proteins specified by tumor suppressor genes stop or suppress cells from dividing when they shouldn't. A disastrous situation arises when a cell that already has mutant proto-oncogenes also gets mutations in tumor suppressor genes. Without brakes, there is nothing to stop a cell with mutant proto-oncogenes from nonstop divisions, leading to a gang of identical cells that also keep dividing. Out of the trillions of cells in a person's body, it only takes one with mutations in a few proto-oncogenes and tumor suppressor genes to start a tumor that can doom all the other cells if not destroyed.

For Leonard Betts's cells to become cancer cells, both versions of at least one important tumor suppressor gene are probably mutated—most likely the gene with the unusual name of p53. p53 is named for the gene's protein product (the "p" stands for protein, and "53" stands for the weight of that protein in tiny kilodalton units[1]).

---

[1]The dalton, also known as an atomic mass unit or amu, is the basic unit of mass on the atomic scale. One dalton is defined as one-twelfth the atomic weight of carbon. A kilodalton is 1,000 daltons, and p53 therefore has the same mass as approximately 4,400 carbon atoms.

The p53 protein is so important that it is known as the guardian of the genome—the brakes in the cellular automobile. When you understand how the p53 protein performs its guardianship duties, it becomes clear why mutations in the p53 gene accelerate cells down a one-way path to disaster.

The duty of the p53 protein is to keep mutations out of a cell's DNA. It only takes a single rogue cell to become a tumor, so keeping your DNA free from mutations that just might strike important genes is clearly in your best interest. p53 can't replace the sunscreen lotion so important in shielding your cells from the sun's radiation, nor can it stop you from inhaling DNA-damaging chemicals from smoking cigarettes. What the p53 protein can do is monitor the DNA of a cell for damage due to chemicals or radiation. If p53 senses that the DNA has been harmed, it literally stops the cell in its tracks and refuses to let it divide until the cell has repaired its damaged DNA. If injury to the DNA of a cell is extensive, then p53 causes the cell to commit suicide before it can become a cancer cell and harm the body.

When the p53 guardian is absent from a cell because random mutations just happen to hit both versions of the p53 gene, the consequences are severe. No longer will that cell stop dividing long enough to repair any damaged DNA and no death awaits a severely compromised cell. Mutations begin accumulating throughout the genome of the cell because the cell doesn't stop to repair the damage. The more mutations that strike the genome, the more chances that some of them will land in proto-oncogenes or tumor suppressor genes. It's no wonder that the cancer cells in about half of all tumors have mutations in both versions of the p53 gene.

If many of Leonard Betts's cells are truly cancerous as Scully and the pathologist believe, then examination of Leonard's DNA would probably reveal that he has inherited two faulty versions of the p53 gene. Normally, this condition is fatal. Mice without at least one normal version of the p53 gene in every cell die within a few weeks of birth riddled with tumors. Based on the mice studies, only fictional people like Leonard could survive with two faulty versions of the p53 gene in every cell.

If an early death awaits an individual who inherits two bad versions of the p53 gene, what are the consequences of inheriting one faulty version of the gene? This would mean that every cell in a person's body has the distinct disadvantage of starting life with only one good version of their genome guardian. While a single version of the gene produces enough p53 protein to guard a cell, all cells are only a single mutation in the remaining good version of the p53 gene away from disaster. The odds that some cell of that person's body will be struck by a mutation in the remaining good version of the p53 gene are much higher than the odds that two independent events will knock out both p53 versions in a cell of a normal person. If you win the lottery, the odds are much greater that you will win again before your jealous neighbor wins twice. If you have just one normal version of the p53 gene in your cells, then any of the 100 trillion cells in your body need only a single mutation in that remaining gene to lose its genome guardian.

In 1969, F. P. Li and Joseph Fraumeni were studying the genealogy of families that had many members afflicted with cancer at an early age. They discovered that a predisposition to cancer could be inherited. In 1990, the nature of that inheritance was made clear. Seventy percent of the families ravaged by a wide variety of cancers were passing along a faulty version of the p53 gene. Inheriting a bad version of the p53 gene means a 50 percent chance of developing cancer by age thirty and a 90 percent chance by age sixty. This particular predisposition to cancer was named Li-Fraumeni syndrome and afflicts one hundred known families. Inheriting mutations in other tumor suppressor genes is what predisposes additional people to more specific types of cancer, such as colon or breast cancer.

At the end of the episode, Leonard Betts traps Scully in an ambulance but she deftly turns the tables by killing him with a defibrillation machine. Apparently, Leonard's regeneration skills are not sufficient to repair the damage to his heart before it stops functioning. Before he dies, Leonard uses his uncanny—and unexplained—ability to detect cancer in others to tell a shocked Scully that she has cancer. And, unfortunately for Scully, it is not the type that allows regeneration of a diseased body.[2]

[2]Scully's cancer treatment lies ahead in Chapter 4.

Aliens 'R Us

INT. KENSINGTON BUILDING—NIGHT

STRUGHOLD

We've been forced to reassess our role in colonization by new facts of biology which have presented themselves.

GROUP ELDER
(speaking up)

The virus has mutated.

WELL-MANICURED MAN

On its own?

CIGARETTE SMOKING MAN

We don't know. So far, there's only the isolated case in Dallas.

STRUGHOLD

Its effect on the host has changed. The virus no longer just invades the brain as a controlling organism. It's developed a way to modify the host body.

WELL-MANICURED MAN

Into what?

STRUGHOLD

A new extraterrestrial biological entity.

—Fight the Future (the X-Files movie)

As those who have seen the *X-Files* movie already know, two Neanderthal explorers had an alien encounter in snow-covered Texas about 37,000 years ago. Since Neanderthals mainly confined themselves to Europe and western Asia, these hominids must have been the Christopher Columbuses of their generation. While exploring a large cave, they find a dead comrade frozen in the icy walls. The corpse's killer soon makes an appearance, attacking one of the unsuspecting Neanderthals with feral glee. The Neanderthal, no easy victim, slashes the creature with a knife before being overwhelmed. His companion dispatches the weakened creature, which leaks a black, oily substance onto the floor of the cave. The remaining Neanderthal has only seconds to savor his victory before being invaded by the black slime.

The underground cave is rediscovered in modern times when a Texas boy accidentally falls through a hole in the roof. Both the kid and his firemen rescuers are overcome by the black slime, still kicking after all these years. The effect of the ancient black slime on humans is not a pretty sight. It causes a lizardlike entity to develop inside the body, the same Neandercidal creature from prehistory. When the entity finishes developing, it bursts through the chest of the victim into the outside world, and indiscriminately attacks anyone nearby.

Chris Carter, the movie's scriptwriter, originally wanted to connect the black oily slime and the lizardlike entity in the following way: the black slime carries a virus, which is infectious when the slime enters a person; in the warmth of a person's body, the virus develops into the monstrous creature.

Very imaginative, yes. Minutely possible within the framework of biology, no.

After reading the movie script in early 1997, I hoped that Chris would change his mind. Having a special place in my heart for viruses, I discussed with him why a virus couldn't possibly develop into anything. Viruses are, after all, just a bag of genes. A bag of genes that turns into a lizard with large black eyes and long pointed nails wouldn't fit even my expanded definition of an extraterrestrial virus. I explored with Chris an idea for tinkering with his scenario. What if the black slime virus is responsible for the development of the creature but is not the progenitor of the creature? The virus, carried into a human by the black slime, could invade a cell in the person's body and cause the cell to lose its identity. The cell could then be enticed by the virus to enter a new developmental pathway. That cell, together with its descendant cells, would regenerate into the hideous alien monster. I was thrilled that Chris liked the changes, since I wasn't enthusiastic about the ribbing I would have taken from my fellow virologists if viruses changed into lizardlike aliens on the big screen—with me credited as science advisor.

I must confess that the idea of a foreign organism invading a living creature and redirecting the identity of some of its cells is not exactly original. A common soil bacterium named *Agrobacterium tumefaciens* sets up shop in plants after persuading the plant to build it a comfy home. The allure of the bacterium is its remarkable and unique ability to remove some of its own DNA and splice it into the genome of a plant cell. The genes on the DNA that the bacterium transfers into the plant cell cause the cell to become confused and lose its identity. The cell then begins rapid cell divisions, generating a mass of cells called a gall. While the gall doesn't continue developing into a homicidal plant monster, it does cause an unsightly lump to appear on the plant, a lump the bacteria call home. The transferred DNA also literally supplies the icing on the cake, since it causes the cells of the gall to prepare daily nutritious repasts for its tiny residents.

Knowing that *Agrobacterium tumefaciens* has the ability to redirect the identity of cells, it's not a stretch to imagine a virus that can do the same. In fact, there are a number of viruses that can insert themselves into the genomes of host cells and then pop back out. Some of these viruses carry with them mutated versions of normal cellular genes, which they stole long ago when extracting themselves out of some other cell's genome. Many of these kidnapped genes were once cellular proto-oncogenes, the precancer genes that when defective can start cells down the road to cancer. When the virus inserts itself and its stolen genes into the DNA of a cell, the stolen proto-oncogene becomes an oncogene and causes the invaded cell to begin nonstop divisions. A cell that is dividing needs to make a number of ingredients to replicate its own DNA, which are the same ingredients that the clever virus needs to duplicate itself as well.

These viruses, called tumor viruses, merely switch a cell from normal mode into cancer mode. The virus carried by the black slime has a much more demanding job. Not only does it have to cause an infected cell to become confused and lose its identity, but the virus then has to make the cell develop into something that isn't human. This would require the presence and activation of an entirely new set of genetic instructions in the human genome—instructions beyond the ones normally used to create a living, breathing, thinking person.

If the science fiction scenario of a virus activating a resident program in the genome of a human cell to cause the cell to develop into a monster is to be even remotely believable, there would have to exist in the human genome an extensive amount of DNA whose purpose is not currently known. One would think this unlikely, since after millions of years of evolution, the human genome should stand out as a shining example of the efficient use of DNA. In contrast, bacteria, being merely single-celled organisms many of which have survived virtually unchanged for over a billion years of evolutionary history, should be able to blame their inefficient genomes for their lowly status on the totem pole of life. There is just one little problem with these last two statements—the opposite is true. Bacterial genomes are highly efficient assemblies of wall-to-wall genes. In contrast, the 100,000 genes of the human genome take up less than 5 percent of the DNA in a cell. That leaves 95 percent of the human genome for . . . ?

Like all humans, your 100,000 genes turn on and off in different cells and at different times during your life. Only when genes are on and active can proteins be made from the instructions specified by the genes. For years, scientists focused on understanding the genes of an organism since it's the proteins that make us who we are. The remaining 95 percent of the human genome is known as "junk" DNA, which pretty much sums up its importance to most scientists. Although little time and money has been spent on understanding why the human genome is filled with junk, scientists have determined that much of the garbage DNA is composed of thousands or even millions of copies of tiny repeated nucleotide sequences, some as short as two nucleotides. About 1 percent of the genome is composed of remnants of past viral infections—pieces of viruses that are now trapped in the genome and can no longer pop back out. Also included in the junkyard of the genome is a small segment that is found in half a million copies scattered throughout every chromosome. These small segments, called Alu, create new versions of themselves as they hop around the genome. If an Alu segment jumps into a spot already occupied by an important gene, that gene becomes disrupted and nonfunctional. The mystery of junk DNA is this: If it's just taking up space, or is potentially harmful to important genes, why exist at all?

The very presence of junk DNA after millions of years of evolution suggests that "junk" is a misnomer. The DNA is probably there for a reason, we just don't know what that reason is . . . yet. Maybe it helps to control when genes are turned on and off, or maybe it helps to maintain the structure of chromosomes. Or maybe it's a set of instructions for the development of a lizardlike alien monster with sharp, pointed teeth.

In "The End," the *X-Files* episode that precedes the movie, a link is first presented between the human genome and the alien conspiracy that permeates many earlier episodes. A twelve-year-old boy named Gibson Praise has become a chess prodigy because he can read the minds of his opponents. Mulder is told by a man who was sent to kill Gibson that the kid is the missing link in Mulder's quest for the truth behind the paranormal events in the X-Files.

Mulder comes to believe that Gibson is the key to all spiritual and paranormal phenomena. He bases this belief on intriguing research done by Scully. She claims that neurological exams of the boy's brain reveal something peculiar in the part of the brain she calls "the God module"—a portion of the brain that may hold the secret to highly emotional religious experiences.

All manners of behaviors are now being attributed to genes. If you are shy or extroverted, intuitive or pessimistic, thrill-seeking or aggressive, chances are that you can blame your genes more than your environment. Only a few decades ago, psychologists were convinced that the environment was the main shaper of our personalities. What changed their minds were studies of twins, many of whom had been separated soon after birth. The personalities of thousands of identical twins, with their identical genomes, were compared with the personalities of fraternal twins in a famous study by the Minnesota Center for Twin and Adoption Research. Their findings? Genes determined personalities more than the environment. In a complete turnaround from earlier thinking, environment is now thought to make family members more different, rather than making them more the same.

The most interesting case to arise from this study was a pair of separated identical twins named Jim. When they were reunited in

1979, both stood six feet tall and weighed 180 pounds. They walked with the same gait and had the same gestures. Each married and divorced women named Linda and then married women named Betty. Chevrolets were their cars, and Miller Lite was their beer. They were addicted to Salem cigarettes, bit their nails, loved stock car racing, and hated baseball. These astonishing coincidences could make someone wonder which gene in the human genome makes a person like women named Betty.

Now there is evidence that religious and paranormal experiences may also be a function of genes. A group of scientists led by Vilaynur Ramachandran at the University of California at San Diego is tiptoeing around the idea that people have vivid religious experiences because of the neural machinery in the temporal lobe of their brains.

Ramachandran studies people with temporal lobe epilepsy, who experience intense religious ecstasy, a sense of "seeing God" and being one with the universe during seizures. Ramachandran believes that there is a region of the temporal lobe of the brain—termed "the God module"—which has neuronal connections that are dedicated to religion and a belief in God. It's this area of the brain that is stimulated in people with temporal lobe epilepsy when they hear religious words. These neuronal connections could have evolved over evolutionary time to encourage loyalty to a tribe, reinforce ties of kinship, or impose order and stability on the social structure. Ramachandran carefully points out that his findings do not deny the validity of religious experiences. He is only pointing to a section of the brain that is stimulated during these experiences. How someone interprets Ramachandran's results depends, as he puts it, on whether one believes that God created the mind, or the mind created God.

The temporal lobe where the putative "God module" resides is a fascinating area of the brain located just below the cerebrum. Sometimes when the temporal lobe is damaged by a stroke, people lose the ability to recognize faces. There is no loss of vision, no problems with language or reading, "face" simply stops being a useful concept in distinguishing one person from another. A second, very rare disorder involving the temporal lobe and face recognition is Capgras syndrome. People with this syndrome insist that parents, spouse, siblings, friends, and pets have been replaced with identical imposters. Some people suffering from this syndrome think that their loved ones are imposters only when they are looking at them, not when speaking to them on the telephone.

Scully tells Mulder that there is tremendous neuronal activity in the God module area of Gibson's brain, which allows him to read minds. This leads Mulder to point to Gibson as genetic proof that paranormal X-File cases can be explained scientifically. The proof, Mulder thinks, is hidden in Gibson's genome—in portions that are unused by normal people but active in Gibson. Unfortunately, before Scully can provide further evidence for Mulder's beliefs, Gibson is kidnapped and the X-Files go up in flames.

In "The Beginning," the opening episode of the sixth season, Mulder is convinced that there is a connection between Gibson and the virus involved in activating the development of the lizardlike creatures introduced in *Fight the Future*, the *X-Files* movie. Unfortunately, hard evidence is lacking. Gibson is gone and the creatures have vanished. Mulder's only piece of evidence is a trace amount of the virus that he believes is extraterrestrial. Mulder is crushed when Scully cannot support scientifically the alien nature of the virus. After running tests, Scully reveals that the virus, while of an unknown type, has the same four nucleotides in its DNA and the same twenty amino acids in its proteins as earthly viruses. She therefore concludes that Mulder is mistaken. The virus comes from Earth.

Mulder's search for hard evidence to back up his beliefs grows warmer when he learns that another alien creature is on the loose. A scientist in Arizona working to develop an antidote and vaccine for the virus accidentally pokes himself with an infected needle. Twelve hours later the creature is born, leaving a claw in the wall of his house and a gaping hole in the chest of the dead scientist. While this might seem like an extreme way to "give birth," it does not necessarily lend support to an alien origin of the creature and virus. A worm named *Caenorhabditis elegans* normally doesn't have much problem giving birth to baby worms. However, a mutation in the wrong worm gene, and giving birth no longer is an option. The mutant worm, not being too bright, doesn't realize the dangers of bearing young until the little nippers blow up Mom when they get too big.

Mulder searches for the entity in hopes of using it to bolster his claims of the existence of extraterrestrials and, of course, prevent it from killing any more people. The creature heads for a nuclear power plant, leaving Mulder to suggest that warmth is needed because the creature is still developing. Mulder comes tantalizingly close to the entity, which has violently attacked several others. In the reactor room of the plant, Mulder finds Gibson, who is helping his kidnappers locate the creature. Blocked by a locked door, Mulder can only stare in horror as the creature slices and dices the kidnapper and then disappears with Gibson. Soon afterward, the creature molts into a familiar-looking gray alien.

At the end of the episode, Scully makes one last attempt to convince Mulder not to ignore her science. She presents him with the DNA test results on Gibson Praise, the virus, and the claw left behind by the creature in the house of the dead scientist. The test that Scully performs is one that allows her to look for copies of the virus DNA in the genomes of Gibson and the entity. What she finds shocks her. Gibson and the creature have the virus DNA in their genomes. When Mulder asks if they are infected with the virus, Scully explains that it is much more than a simple infection. The viral DNA in their genomes is a genetic remnant—it has been in the human and creature genomes for a very long time. But there is more. The virus DNA is not only in the genomes of Gibson and the creature, it is in everyone's genome. But the virus DNA is no longer functional. Where once it could be activated, it now makes up part of a person's inactive junk DNA. Gibson, however, is the exception. In his genome, the virus DNA is active—its genes are turned on.

```
            INT. X-FILES OFFICE—DAY


                    MULDER
        I can't accept that. Not if it refutes what I know is true.


                    SCULLY
        They're test results. DNA from the fingernail we found—
        matching exactly DNA in the virus you believe is extrater-
        restrial—


                    MULDER
                    (realizing)
        —that's the connection—
```

                            SCULLY
                           (continuing)
        —which also matches exactly DNA I found in Gibson Praise.

Mulder takes a moment. Because it changes everything.

                            MULDER
        I don't understand. He's infected with the virus?

                            SCULLY
                          (dead serious)
        No. It's part of his DNA. In fact, it's part of all of our DNA.
        It's called a genetic remnant. Inactive junk DNA. Except in
        Gibson it's turned on.

                            MULDER
        If it's true, it'd mean the boy is in some part extraterres-
        trial.

                            SCULLY
        It would mean we all are.

                                        —"The Beginning"

4
Releasing the Genetic Genie

Introduction

EXT. ALLEY BEHIND MOTEL—NIGHT

> **DEEP THROAT**
> In the early fifties, at the height of the cold war, we got wind that the Russians were fooling around with eugenics. Rather primitively, I might add—trying to crossbreed their top scientists, athletes, you name it, in order to come up with a superior soldier.
> > (beat)
> Naturally, we jumped on the bandwagon...

> **MULDER**
> These "Litchfield" experiments...

> **DEEP THROAT**
> > (nodding)
> A group of genetically controlled children were raised and monitored on a compound in Litchfield. The boys were all called Adam, and the girls were all called Eve.

> —"Eve"

In December 1996, five hundred earnest members of the freshman class listened with nearly rapt attention to my final lecture after a long semester of Introductory Biology. The subject of the day was developmental biology—how a complex organism with trillions of cells is formed from a single fertilized egg cell. The culmination of the course and the lecture was a special topic:

Why cloning mammals from adult cells is impossible.

To make my point, I carefully interwove pieces of the students' newly acquired knowledge about DNA and genetics into a story of how cells become specialized as the embryo develops. I explained how some of the DNA in specialized cells becomes irrevocably altered by the attachment of special chemicals and proteins—attachments that permanently shut down genes that are no longer necessary. I added that such irreversible alterations to the genes of mature cells means that the complete DNA information packet in the cell is no longer available to reproduce an entire organism; hence cloning a new organism from the DNA of adult cells is impossible.

It was a performance that the students deemed worthy of a standing ovation. The semester was over and I was satisfied knowing that five hundred budding biologists were venturing out into the world with a proper appreciation for the nature of cells and molecules.

Three months later, I picked up my weekly issue of the journal *Nature* and was stunned to see a picture of Dolly, the most famous sheep in history by virtue of being the first lamb cloned using DNA of an adult sheep cell. My initial thought after recovering from shock was that the final lecture of Introductory Biology was going to need a little tinkering.

It's rare when a single experiment dramatically changes a fundamental tenet of biology. In a single year, every textbook on molecular, cellular, and developmental biology was suddenly out of date. In a single year, I went from lecturing about the impossibility of cloning to giving my students daily updates on the births of the first cloned cows—the products of ten years of persistence by Dr. James Robl, a friend and colleague at the University of Massachusetts. A half year after the cows were born, twenty-two tiny Hawaiian mice—three generations of identical animals—joined the ranks of the cloned. Stories on animal clones were front-page news, and their cloners became reluctant celebrities. The media's focus on cloning will reach a new high with the expected announcement of cloned monkeys within the next year or so—until the inevitable confirmed cloning of the first human.

A clone is a genetic duplicate of something, whether a gene, cell, cow, or person. Many biology laboratories are engaged daily in some type of cloning. In my own lab, we clone bacteria every time we entice one to divide and produce two identical bacteria. We clone viruses when we use the technique of polymerase chain reaction (PCR) to make millions of copies of a virus's genetic material. Ever take a cutting of a plant and use the cutting to grow a second plant identical to the first? Then welcome to the ranks of the cloners. Plant clones, however, don't seem to have the same appeal as the current parade of animal clones. The reason, of course, is obvious. If mammals can be cloned—sheep, cows, mice, and probably many more by the time you read this paragraph—why not people? Why not you?

If your information on cloning humans or other mammals comes mainly from movies like *Sleeper* (1973), *Multiplicity* (1996), or *Alien Resurrection* (1997), you probably have a less than realistic understanding of the subject. In *Sleeper,* the nose of the person to be cloned is placed on a table shaped in the form of a man, a few dials are turned, and a man should appear a few hours later ready to continue where he left off. Cloning in *Multiplicity* requires the use of a giant, three-dimensional duplicating machine. Scan the original, take some red blood cells, and presto, a fully formed clone is born, all memories intact.[1] Reproducing people by the *Sleeper* or *Multiplicity* methods has about as much chance of succeeding as duplicating your computer by laying the Pentium chip on top of a copy machine and pushing the green button. In *Alien Resurrection,* blood cells are used to clone Ripley (Sigourney Weaver), memories nearly intact, along with the toothy alien parasite lodged in her chest. One can only wonder if her last meal was "cloned" as well.

The most common misconception about cloning is that the

---

[1]The red blood cells are a curious cloning ingredient, since they are the only cells in the body that lack DNA . . . and DNA is the most important ingredient in cloning.

clone ends up being the same age as the person being cloned. As accurately portrayed in the Hitler-cloning thriller *The Boys from Brazil* (1978), cloning an organism does not replace the normal stages of development. A cloned organism must form inside a surrogate mother and be born a baby, whether a lamb, calf, mouse, or maybe, someday, human. Memories are not cloned. Stomach contents are not cloned. It's simply like having an identical twin born many years after the first twin. A baby twin who doesn't have a clue about what the clonee has been up to all these years.

To better understand cloning, let's say that you want to clone a dear friend, your cherished cow, Bessie, who has one hoof in the grave. What you will end up with at the end of the cloning process is a calf with the same genetic material, the same DNA genome, as your beloved Bessie. All of Bessie's cells (except for her red blood cells and some immune system cells) have a complete DNA genome, so each cell contains the instructions for making another Bessie. But you need more than just Bessie's cells to make a clone. If you place an ordinary cell, such as one of Bessie's skin cells, in a culture dish and bathe it in a nutritious liquid, the cell will divide and give you more skin cells. What won't end up mooing in the culture dish is a baby Bessie, with her two hundred different cell types. Trying to convince one of Bessie's cells to make you a cow all by itself is like placing a set of computer schematics on a table and expecting a new computer to materialize around them. While the DNA of an adult cow cell may contain all the information needed to re-create Bessie, the rest of the adult cell is incapable of running the create-a-cow program.

What you need is an egg . . . not necessarily one of Bessie's. Any unfertilized cow egg will do. Eggs are experts at running the organism development program. An egg is able to divide rapidly as soon as it is fertilized with a sperm because it has stored ahead of time all the ingredients required to undergo cell division. A fertilized egg needs only to duplicate its DNA genome, which is a combination of the half genomes of the egg and sperm, before it can divide. After the fertilized egg has divided into two cells, the new cells still have enough material stored inside to continue dividing quickly, the pace of divisions not slowing down until all of the stored material is used up.

To clone a baby Bessie, you therefore need an unfertilized cow egg and the DNA genome from one of Bessie's cells. The reason it doesn't matter what cow you get the egg from is that one of your first steps will be to remove the genome—or, rather, the half genome contained in every egg. This is done either by UV or laser irradiation or by deftly yanking the chromosomes out of the egg. Now you have an egg without genetic material that is prepared to host Bessie's DNA genome. Getting Bessie's DNA into the egg requires some intricate microscopic manipulations. The DNA in Bessie's cells reside in the little nucleus compartment. The nucleus needs to be removed from one of Bessie's cells carefully so that it doesn't pop and spew chromosomes all over the counter. After the nucleus is placed in a tiny needle, it is injected into the prepared egg and the egg is ready to begin the development process. Another method that some cloners use is called cell fusion. The egg cell and one of Bessie's cells are fused together with a jolt of electricity to become one cell. After fusion, the nucleus from Bessie's cell becomes the new nucleus of the egg cell, ready to initiate the create-a-cow program.

Whichever method you choose, the result is an egg that contains Bessie's DNA genome. What happens next is shrouded in mystery. Somehow, unknown components in the egg cause the adult DNA genome to restore itself back to a condition before it was part of a specialized adult cell. The chemicals and proteins attached to some of the genes in the adult DNA are removed by an unknown mechanism, which restores the DNA to the pristine condition necessary to initiate the organism development program. Once this occurs, the egg with its newly reset DNA starts to divide. When a small ball of cells has formed, the developing embryo can be implanted inside a surrogate mother cow. Then you can sit back, relax, and wait a little over nine months for the happy event.

But will it be a happy event? If you are expecting an exact duplicate of Bessie, personality and all, you will likely be disappointed. Just as identical twins have different personalities while sharing identical DNA genomes, so will the calf and Bessie. The cloned calf and Bessie will have different birth mothers, whose ideas on nurturing the infants might not coincide. In fact, some behavioral scientists are interested in cloning animals to determine what role the birth mother plays on the personality of the baby. The barnyard playmates will also be different, which may impact on baby Bessie's personality.

The cloned calf and Bessie are also less identical than identical twins. This is because the nucleus of a cell doesn't contain the only DNA in the cell. Outside the nucleus, floating in the main compartment of the cell, are little cubbyholes called mitochondria, the remnants of symbiotic prokaryotic cells. Mitochondria have their own tiny DNA genomes that specify thirteen proteins, including some important enzymes involved in energy production. Bessie only provided the nuclear DNA for the clone and not the mitochondrial DNA, which came from the donor egg. Therefore, the clone will only be a perfect Bessie duplicate if Bessie herself supplied the egg.

As you get together the materials and equipment for the Bessie cloning project, you begin to have some additional thoughts. Bessie is a wonderful cow, to be sure, but couldn't she be a bit better? Bessie, with all her endearing qualities, is a bit shorter than the cow ideal. And then Bessie, being white, tends to show the dirt. What if you could clone not only Bessie, but a new and improved Bessie? A bigger, blacker Bessie?

If you want a better Bessie, improvements need to be made before Bessie's DNA is added to the egg. If you want to make Bessie bigger and blacker, you will need to change her genome, a process called genetic engineering. Making Bessie bigger will require adding extra copies of the gene that specifies a protein called bovine growth hormone, which is normally produced in the pituitary gland and makes people and animals grow larger. To make Bessie black, you need to add a gene that specifies black coat color.

For many years now, scientists have been dissecting the genomes of organisms from bacteria to man, searching for genes responsible for diseases or genes that confer various traits. Segments of DNA containing all sorts of different genes have been separated from the rest of the DNA in an organism's genome so that the individual genes can be studied in the lab. While the cow gene for growth hormone was isolated many years ago, the cow genes for hair color are still hidden in the cow genome. Isolating individual genes can take many years and hundreds of thousands of dollars, a considerable expense just to make a cow black. There is, however, a cheap alternative that might work. The black hair color gene has been isolated from mice, and mice are mammals just like cows. If you add the mouse black hair color gene to a cow's genome, maybe it will turn the cow black. At least, you don't know until you try.

If you don't have handy copies of the cow growth hormone gene and the mouse black hair color gene in your freezer, you need to contact people who do have the genes. Luckily, the scientific community is filled with altruistic people. There are web sites that let you know what genes of an organism have been isolated and who did the isolating. Simply check out the bovine genome project web page (http://www.ri.bbsrc.ac.uk/bovmap/bovmap.html) and the mouse genome project web page (http://www2.igh.cnrs.fr/Mouse-GenomeDBS.html). It's entirely up to you whether you choose to grovel on the telephone or by e-mail to obtain copies of the gene.

To genetically engineer your cow clone, you need to add the growth hormone gene and the black hair color gene to the genome of one of Bessie's cells before you transfer her DNA into the egg to make the clone. Methods for adding DNA to mammalian cells have been around for twenty-five years so this step isn't a problem. DNA segments can be injected into cells, or cells can be shocked into taking up DNA added to the outside of a cell. But as you prepare to add the two genes to some of Bessie's cells, you can't help feeling a bit selfish that so much time and energy is going into making a cow that only you will love. If you are going to go through all the trouble of making Bessie bigger and blacker, why not make a cow that the whole world will appreciate, a cow that produces something in her milk that will benefit mankind?

You therefore decide to add one more gene to Bessie's cells. Emulating the scientists at PPL Therapeutics in Scotland, who produced both Dolly and the first genetically engineered sheep clones, you decide to add the gene for human clotting factor IX, a protein required by people suffering from hemophilia so that their blood will clot properly. Currently, hemophiliacs must obtain factor IX from human blood donors, which puts them at risk for blood-transmitted diseases like AIDS and hepatitis. Getting the clotting factor from sheep or cow milk will be safer and cheaper, to say nothing of tastier. With these altruistic thoughts in mind, you add the human factor IX gene along with the growth hormone gene and black hair color gene to the genome of one of Bessie's cells. Then you transfer the nucleus of that cell into the unfertilized cow egg, let the cells divide a bit, and then implant the embryo into a surrogate mother cow.

As the little Bessie clone starts developing inside her surrogate mother, another idea comes to mind. Why stop at just one Bessie? Why not whole herds of beautiful Bessies? The developing fetus is a treasure trove of new and improved baby Bessie cells. So just like Jim Robl did when he cloned the first cows, you mournfully sacrifice the fetus and separate the cells from each other. In the nucleus of each cell is the genetic material for producing a new and improved Bessie, and you have thousands of cells. By placing the cells in culture dishes, and enticing the cells to divide, you can grow even more cells—dishes and dishes of improved Bessie cells. The nucleus of each cell can be merged with an unfertilized egg and implanted into a surrogate mother. Soon, thousands of big, black beautiful Bessie calves are born. Ordinary people are so enchanted with the delightful cows and hemophiliacs worldwide so appreciative for making their lives less stressful that a grateful public urges that you be awarded the Nobel Peace Prize . . . making you gaze a little differently at your gorgeous goat, Bob.

Cloning a pet is probably not too many years down the road. In a recent *USA Today* Internet poll, about 65 percent of respondents said that they would clone a pet if they had the money. Right now, if the pet is a sheep, cow, or mouse, a clone could probably be produced at the bargain-basement cost of about $4,000, providing you can convince a scientist who would rather be doing important research to clone your pet instead. Cow-cloner Jim Robl told me that he has been deluged with requests from pet owners to clone their beloved dog/cat/horse. The letters come complete with pictures of the pet and descriptions extolling how their dog/cat/horse is the world's most unusual, most intelligent animal. The letters describe the heroic deeds of their dog/cat/horse—how it jumped into a swimming pool and saved a precious dog/cat/mouse/child from certain death. But knowing how to clone a cow doesn't translate into knowing how to clone other animals. The method that works for one type of animal doesn't seem to work for others. Any pet cloning "industry" will need to develop unique procedures for each type of animal, and working out a method is very expensive. Still, this doesn't deter some people. One millionaire is funding a Texas A&M lab with up to 5 million dollars to clone a much-loved dog, Missy.

Cloning is just one of the consequences of the genetic revolution. A better understanding of genetics has allowed the genomes of organisms to be subtly altered—plants can be genetically engineered to make vaccines, and animals can produce pharmaceuticals in their milk and urine. There is even a name given to this new industry: pharming. Understanding the nature of genes and the underlying causes of diseases like cancer and cystic fibrosis is leading to treatments known as gene therapy—adding good genes to cells of people to counteract the bad. What was once science fiction has become science reality.

Genetic engineering, gene therapy, and cloning are no strangers to *The X-Files*. Scully's nasopharyngeal cancer is treated with the latest in gene therapy, at least, that's what the doctor says. In "Eve," genetically engineered clones named Eve aren't precisely the superwomen that their creators had in mind. Copies of Mulder's sister, Samantha, have popped up in so many episodes that Mulder (and we) may never meet the original. And in "Herrenvolk," plants genetically engineered so that their pollen contains virus is part of a decades-long plot to colonize the planet by creatures who do not wish us well. In *The X-Files*, as well as in real life, the genetic genie is a tool that can be used to cause both harm and good, either knowingly or unknowingly

Take Two Genes and Call Me in the Morning

INT. MRI LAB—HOLY CROSS MEMORIAL HOSPITAL

SCULLY

The truth is...the type and placement of the tumor make it difficult. To the extreme.

MULDER

I refuse to believe that.

She takes his hand, holds it tight.

SCULLY

For all the times I've said that to you—I'm as certain about this as you've ever been.

(beat)

I have cancer. The mass is on the wall between my sinus and
cerebrum. If it pushes into my brain, statistically there's
about zero chance of survival.

MULDER
You can't stand here and tell me there's nothing—
(refusing her)
There must be people who've received treatment—

SCULLY
(tentatively)
Yes, there are...

—"Memento Mori"

At the end of the episode "Leonard Betts," Scully is shocked to
learn that she has become a victim of cancer. In "Memento Mori,"
Scully discovers that her cancer is a rare nasopharyngeal mass, located on the wall between her sinus and cerebrum. Although Scully
doesn't have the normal warning signs of this type of cancer—trouble breathing and speaking or a ringing in the ears—the magnetic
resonance imaging X rays leave no doubt as to the diagnosis.

Scully, as a medical doctor, is surprised that she has this particular type of cancer. Nasopharyngeal tumors are rarely found in North
Americans or Europeans, but are relatively common in Southern
Chinese and Eskimos. A higher frequency for a certain type of cancer
in a particular population usually implies genetic and environmental
risk factors. Nasopharyngeal cancer is associated with eating salted
fish and preserved foods, as well as smoking. More recently, scientists
have found an ominous intruder inside the tumor cells of everyone
who has nasopharyngeal cancer: the Epstein-Barr virus.

If you ever need a sample of the Epstein-Barr virus, look no farther than your own body. Nearly everyone is or was infected with
this virus. If your infection occurred during childhood, there were
few if any symptoms. If you caught the virus as a teenager, you probably experienced the joys of mononucleosis. Mononucleosis is
known as the "kissing disease" because Epstein-Barr virus is transmitted between people through saliva, which usually occurs when

kissing. While the virus remains dormant in most infected people, it can in rare cases become activated, which can lead to nasopharyngeal cancer. Epstein-Barr virus is also associated with Burkitt's lymphoma, a tumor found in children from equatorial Africa.

Scully's nasopharyngeal cancer was probably not caused by infection with the Epstein-Barr virus while eating copious amounts of preserved salted fish between cigarette puffs on a long-term holiday in Alaska. Acquiring such a rare cancer leads Scully to believe a link exists between her cancer and her abduction several years earlier. To follow up on this theory, Mulder and Scully visit a group of women who experienced similar abductions and had talked to Scully previously. When Mulder and Scully arrive at the home of the group's organizer, Betty Hagopian, they are stunned and saddened to learn that Betty has recently passed away from nasopharyngeal cancer. And not just Betty. Of the original women in the group, only one remains alive—and she is being treated for the same rare cancer.

Scully finds the woman, Penni Northern, in a hospital suffering the final stages of her terminal cancer. Penni is being treated by a Dr. Scanlon, who tells Scully that she doesn't need to suffer the same fate as the other women provided that she starts treatment immediately. Surgery for Scully's particular nasopharyngeal tumor isn't an option. Dr. Scanlon wants to begin traditional chemotherapy and radiation treatments followed by a new experimental approach, gene therapy. Scully checks herself into the hospital and hopes for the best.

Cancer treatments other than surgery have one simple goal: kill every cancer cell. Let even one cell escape and the cancer will return. What sounds so straightforward has proven to be quite elusive since different cancers respond unpredictably to different treatments. Many scientists have devoted their careers to dreaming up new ways to kill cancer cells without harming normal cells. The dreams all begin with a simple premise: find out what makes cancer cells different from normal cells and exploit those differences to destroy them and only them.

The most basic difference between cancer cells and normal cells is that cancer cells only want to divide and make more cancer cells. Nonstop divisions, though, are not achieved without effort. Cancer cells must work overtime to produce all the necessary ingredients to duplicate their contents for equal distribution to the two daughter cells. Prevent a cancer cell from making the ingredients it needs to divide in two and it will die.

Consider the following analogy. You want to build a house identical to the house next door. One option (not necessarily the best option) is to split the neighbor's house down the center, move the two halves apart, and then duplicate the missing sections for both houses. Wood, glass, piping, furniture, all must be available and fashioned into the proper shape to complete the duplicate houses. Once the process is finished, two identical houses will stand where once there was one—just like two cells are formed from one original cell during cell division. If your town wanted to kill the project, it would have several options. Simply refusing to issue a permit isn't one of them since the order will be ignored—cancer cells don't listen anymore to the cells around them. The town could stop one of the basic ingredients from being available, like wood; no wood, no completing the construction of the two houses. Alternatively, the town could simply jail all the carpenters and plumbers—the materials would be available, but no work would get done. Scientists can stop cancer cells in similar ways, by either preventing the cell from accumulating critical materials or by stopping the actual physical process of splitting in two.

Treating a person with drugs to impede cancer cells from dividing is called chemotherapy. Many drugs have been discovered that block the steps leading up to cell division. One of the most effective anti-cancer drugs, methotrexate, throws a monkey wrench into a cell's ability to make nucleotides, the beads on a DNA chain. If a cell can't make nucleotides, it can't make DNA—and fresh DNA is absolutely required for a cell to divide in two. Another chemotherapy drug, doxorubicin, shuts down one of the enzymes that makes fresh DNA; nucleotides are plentiful, but the cell is unable to string them into DNA. A third popular drug, vinblastine, doesn't stop DNA from being made but does block the cell from going through the final steps of dividing in two. Chemotherapy treatment is usually a combination of these three types of drugs—a deadly cocktail for cells that need to divide to live.

The use of chemotherapy comes with a price. Traditional chemotherapy has been likened to carpet bombing. Instead of a surgical air strike against one building, the entire city is bombed—with a "carpet" of bombs—with the hope that one of the bombs hits the desired target. The reality of chemotherapy is that people get blasted with drug doses just shy of lethal, with the hope that the drugs will wipe out the tumor without wiping out the person. The drugs don't discriminate between cancer cells and healthy cells. Once the drugs are administered, they travel throughout the body getting absorbed by cells that are both normal and diseased. If an agent like methotrexate stops a cancer cell from making nucleotides, it also stops normal cells from making nucleotides.

Fortunately, most of the other cells in the body aren't interested in making nucleotides—they don't need to replicate their DNA since they don't need to divide. If these normal cells can't make nucleotides, so what? They can exist quite happily not making something that they don't need. However, some normal cells do need to divide. Hair cells, bone marrow cells that produce the white blood cells of the immune system, and the cells that line the stomach will die just like cancer cells when the body is treated with chemotherapy drugs. So cancer patients lose their hair, become more susceptible to disease, and are sick to their stomachs during chemotherapy treatment.

While chemotherapy can be very effective against many types of cancer, killing every cancer cell in a tumor is a difficult task. One problem is that cancer cells can mutate and no longer allow the drugs to enter the cell. Cancer patients therefore are frequently treated by several different methods. In Scully's case, Dr. Scanlon wants to try both chemotherapy and radiation therapy.

Radiation therapy is the most common treatment for nasopharyngeal tumors and many other types of cancer. Radiation kills cancer cells in two ways. A high dose of radiation means that a tremendous number of mutations get incorporated into the DNA of a dividing cell. The cell's guardian angel, the protein p53, which normally helps to keep cells from becoming cancer cells, senses that the cell's DNA is highly damaged and then causes the cell to commit suicide for the good of the body. The second way that radiation kills cells is less clear. The p53 protein doesn't function properly in about half of all tumors, yet these tumors still respond to radiation by killing themselves. This leads scientists to believe that there is a second way that radiation kills cells that doesn't involve the p53 protein.

Traditional radiation treatment for a cancer like Scully's, which in "Memento Mori" is still confined to one particular location, involves focusing a tight beam of X rays in the general vicinity of the tumor. Unlike chemotherapy, a whole city isn't bombed. However, normal cells in the neighborhood of the target tumor still receive considerable damage from the imprecise X rays. As Scully is enduring the side effects of her chemotherapy and radiation treatments, she waxes philosophically about her battle against the invaders from within—that destroying cancer comes with a risk of destroying yourself.

For years, scientists have searched for other ways of killing tumor cells while leaving normal cells unscathed. Devising such new treatments requires a great deal of additional knowledge about cancer cells beyond the rather vague "they divide forever." The substantial sums of money spent searching for chinks in the armor of cancer cells is beginning to pay off. Scientists have found that cancer cells have special proteins on their surfaces that are missing from the surfaces of normal cells. It's almost as if the tumor cells paint a bright red X on their outsides and then dare us to figure out ways to attack them.

Many new experimental treatments are designed to do just that. One method involves antibody "smart bombs" that home in on these proteins on the surfaces of cancer cells. Recall that antibodies are simply special proteins that recognize and stick to other proteins like two adjoining puzzle pieces. It's a routine laboratory procedure to isolate particular proteins, like cancer cell surface proteins, and then prepare specific antibodies to these proteins—antibodies designed to latch onto only the cancer proteins on the surface of tumor cells. These antibodies are then made into smart bombs by attaching to them a little present for their target tumor cells—a single radioactive atom. When the smart bombs and their radioactive packages are injected into a patient, they migrate through the body until finding and sticking to cancer cells. The tumor then gets hammered by fifty thousand times the radiation that can safely be applied by a wide beam of X rays from outside the body. Based on the size of the tumor, scientists can choose different radioactive atoms to attach to the antibody. For a small tumor, an atom can be attached that emits radiation extending only a fraction of an inch. For a larger tumor, a different radioactive atom is chosen that emits radiation extending a longer distance. The tumor gets destroyed by the radiation, whose lethal effects drop precipitously beyond a fraction of an inch.

Radioactive antibody smart bombs are just one of the new cancer treatments currently being tested. Dr. Scanlon has chosen a different experimental treatment for Scully: gene therapy. The premise behind gene therapy is simple. If someone has a disease because of a defective gene—and cancer is, after all, caused by defective genes— just add a brand-new, perfect copy of the gene to a person's cells and, voilà, instant cure. Instead of merely treating the symptoms of a disease, gene therapy would correct the underlying problem—a good gene would supersede the presence of a bad gene, and the disease would disappear.

The possibility of gene therapy was science fiction only a few decades ago. Then in 1990, much sooner than nearly everyone thought possible, the age of gene therapy began. Scientists pondered which disease would be the first test case. Cancer was not high on the list since it was far too complex—the corruption of many genes turns a normal cell into a cancer cell. A much simpler disease was needed for the first gene therapy experiment. A disease that could be cured by replacement of a single faulty gene.

There are many genetic diseases caused by single faulty genes—cystic fibrosis, muscular dystrophy, Huntington's disease, to name just a few. Deciding which disease would be the first test for the efficacy of gene therapy required taking into account many considerations. The normal version of the defective disease gene had to be as simple as possible. Most genes have complex on/off switches with built-in instructions for activating and inactivating the gene in different tissues and at different times in the life of an organism. The first candidate gene for gene therapy had to be under simple control, such as "always on" in every cell and at all times. This simplicity was required because there was no guarantee that any complicated switches would work properly if a gene was added to the genome of a cell by gene therapy. Another consideration was how much protein needed to be made by the newly added gene. Since no one could predict how much protein the new gene would cause to be made, a tiny amount of protein or a large amount had to lead to the same result: a cure.

Keeping all these considerations in mind, scientists at the National Institutes of Health selected two little girls, ages four and nine, to be the first recipients of new genes by gene therapy. The girls suffered from a disease called severe combined immunodeficiency, also known as the "boy in the bubble" syndrome. Born with a single defective gene, these children had no natural immune systems. Their faulty gene should have provided all their cells with an enzyme called adenosine deaminase, or ADA for short. Not being able to produce ADA was especially harmful to their immune cells. Without the ADA enzyme, immune cells accumulate a toxic substance and die. All the girls needed to be cured was for scientists to add a good version of the ADA gene to their immune cells. ADA would be produced, the toxic substance would vanish, and the immune cells would live.

The girls had been treated for years with doses of their missing ADA enzyme so that they did have some immune cells. The immune cells needed to be removed from the girls so that the new gene could be added to the cells in a test tube. Fortunately, immune cells are floating around in the blood, and blood is one of those items that isn't very difficult to remove from a person. Once the scientists separated the immune cells from the rest of the blood, the new ADA gene was inserted into the genomes of the cells, and the cells started making the missing enzyme.

The immune cells with their new gene didn't sit for long in the test tube. After growing more of the cells by enticing them to divide, the new-and-improved immune cells were injected back into the little girls. Since immune cells don't last forever, the treatments were continued for several years. In October 1995, the results of this first gene therapy study were published in the prestigious journal *Science*. The verdict? While the girls were not cured, both had significantly improved immune systems. The four-year-old girl, who had been kept isolated in her house, was able to attend a public kindergarten. The headaches and chronic sinus problems of the nine-year-old cleared up a few months after the treatment started. The first gene therapy treatment was considered a success and the age of altering the human genome had begun.

This good news set loose an avalanche of new gene therapy experiments by biotechnology and drug companies designed to cure everything from cystic fibrosis to cancer. Over the next eight years, more than three hundred different gene therapies involving more than three thousand patients were tried worldwide. In the United States alone, 200 million dollars a year is spent by the government on gene therapy with millions of additional dollars invested by biotechnology companies. And the results of all these gene therapy treatments? So far, not a single person has been cured of a disease.

The initial euphoria surrounding gene therapy has been replaced by angry reports stating that the procedure is overhyped and has only led to false hopes and expectations. Gene therapy had so much promise, so why after eight years is that promise so elusive? Finding the problem wasn't difficult: the gene has to get into the right cells—a lot of the right cells. In the case of the two little girls, only their immune cells needed to be altered. This made the process reasonably simple. Immune cells could be removed from their bodies, the gene added to the cells in a test tube, and the cells then returned to their bodies. This isn't an option for most other diseases. To cure cystic fibrosis, the cells of the lungs need to receive the replacement gene. Medical science is not up to the task of removing the lungs, inserting the gene into millions of lung cells, and then replacing the lungs . . . not if the patient's survival is desired. Any gene replacement therapy for diseases like cystic fibrosis must be done in the body. And therein lies the problem.

Cells don't normally put out the welcome mat for foreign DNA. On the contrary, cells have evolved for millions of years expressly to keep foreign DNA out. That's because foreign DNA generally comes in the form of viruses, and virus infections are not in the best interest of a cell. Yet many viruses have found ways of sneaking their DNA into cells. And if viruses can get their own DNA into a cell, then they can sneak other DNA in as well—such as genes for gene therapy. For this reason, the first methods designed to add new genes to cells involve piggybacking the gene onto a virus. The virus genome is opened up and the new gene is spliced in, much in the way that a child disconnects a toy train and splices in a new boxcar. The virus can be thought of as a vehicle; when the virus vehicle drives into a cell, it delivers the gene passenger as well.

But which virus makes the best vehicle? Automobile vehicles are designed with the passenger in mind. Viruses, on the other hand, did not evolve wondering when they were going to be used for gene therapy. No virus is perfect for the task. Take the virus called adenovirus. A terrific vehicle, its proponents claim. It can infect dividing cells. It can infect nondividing cells. It can hold an enormous passenger gene. However, the human immune system is familiar with adenovirus, which normally causes flulike symptoms. When adenovirus carrying genes for gene therapy is injected into a person, the immune system goes on a rampage, causing massive inflammation side effects as it wipes out the virus left and right. If a lucky virus manages to evade the immune system and enter a cell, a second problem arises. The virus doesn't know how to splice itself and its passenger gene into the genome of the cell, which is the only way that the new gene can become a permanent fixture in the cell. Instead the virus and its passenger gene live separate from the genome as an isolated piece of DNA, a living arrangement that is temporary, at best.

To permanently insert a gene into the genome of a cell—a requirement for curing most genetic diseases—a virus vehicle must be able to insert itself into the DNA of a cell so that any passenger gene will also be joined to the genome. Lentiviruses are a group of related viruses that are very good at doing just that. Unfortunately, lentiviruses can only infect dividing cells—good for cancer gene therapy, since cancer cells are busy dividing, but bad for cystic fibrosis and other diseases where nondividing cells like lung cells must be given the gene. And then there is the other little problem about lentivirus vehicles. The lentivirus that makes the best vehicle is a virus called HIV. Yes, that HIV—the virus that causes AIDS. Although the HIV vehicle has been engineered to no longer contain the genes that make the virus harmful, convincing people that they should let themselves be infected with HIV, even a harmless HIV, is going to be a public relations nightmare for biotechnology companies.

So far, there is no perfect virus vehicle, although much effort is going into designing one. But once the perfect vehicle is found (or made), a second major problem remains: how to get the virus with its gene passenger to bypass cells that don't need to be fixed and target the cells that do need fixing. Why is this such a problem? Imagine that you have been given a program on a disk that needs to be inserted into the main computer in a building of a hostile foreign country. In this scenario, you are the lentivirus vehicle and the disk program is the passenger gene that needs to be carried into a cell "building" and inserted into a "computer" genome.

As you parachute into the proper city, you realize that you left the maps in your other suit pocket. Just like a virus, you wander around, aimlessly hoping to run into the proper building, which just happens to look nearly identical to all the other buildings. All this time, the city immune system police have your picture and are looking to eliminate you. If you enter the wrong building, you are trapped and can't get out. Your mission is therefore to avoid the police, enter the correct building, and then deposit the disk into the computer. This is one reason why scientists don't inject a single virus vehicle into someone—they inject hundreds of millions of virus vehicles, each with the same passenger gene. However, even if every single virus vehicle evaded the immune system and every virus infected a different cell, only one out of every fifty thousand cells would receive the new gene—not enough to cure a disease.

There is one final problem to overcome if gene therapy is to become a successful treatment strategy: how to make the virus insert its passenger gene into a reasonably good place in the genome. If the new gene splices itself into the vast junkyard regions of the genome, then the gene probably won't work properly. And if the gene lands on top of an important resident gene, the cell may die or, even worse, start down the road toward becoming a cancer cell.

With all these problems, it is no wonder that gene therapy has had such limited success. Is this then the end of gene therapy? Fortunately, the answer is no. Scientists are figuring out ways for the virus vehicle to stealthily evade the immune system, zero in on the right cells, and insert itself into specific places in the genome. Once these problems are solved, gene therapy should fulfill its destiny as the next great hope of modern medicine.

INT. HOSPITAL PROCEDURE ROOM

A RADIOTHERAPIST steps around the accelerator, angling the head toward Scully, who sits deathly still, in the patient chair.

SCULLY (V.O.)

In med school, I learned that cancer arrives in the body
unannounced—a dark stranger who takes up residence, turning its new home against itself. This is the evil of cancer—
that it starts as an invader, but soon becomes one with the
invaded—forcing you to destroy it, but only at the risk of
destroying yourself.

The Radiotherapist works to secure clamps that will hold Scully's head into
position.

SCULLY (V.O.)

It is science's demon possession—my treatments, science's attempt at exorcism. Mulder, I hope that in these terms you
might know it and know me, and accept this stranger so many
recognize but cannot ever completely cast out.

—"Memento Mori"

Scully, like many real-life patients, doesn't have time for scientists to work out the details. Help is needed now. So gene therapy clinical trials continue using the current level of knowledge. Dr. Scanlon
tells Scully that he wants to treat her cancer with p53 gene therapy.
While this may sound like something that the episode's writers (Chris
Carter, Frank Spotnitz, John Shiban, and Vince Gilligan) pulled out of
a hat, p53 gene therapy is very real. The idea behind gene therapy on
p53 is clever. Recall that one of the reasons why a cell becomes a cancer cell is because it has lost its p53 genome guardian to mutation.
Over one half of all tumors have defective versions of their p53 gene.
One of the roles of p53 is to cause defective cells to kill themselves for
the benefit of the organism. The idea behind p53 gene therapy is to
add a good version of the p53 gene back to tumor cells that have lost
their genome guardian. When the p53 protein is made from the new
gene inside a cancer cell, the protein should be so shocked at the condition of the cell that it should immediately convince the cell to commit suicide. A biotechnology company called Introgen has several
clinical trials under way to test if p53 gene therapy works to shrink tumors, including tumors such as Scully's.

So what is Scully's prognosis following p53 gene therapy? Based on the results of early clinical trials, not bad. Some people with advanced tumors that were recalcitrant to treatment using conventional methods had a partial remission of their cancer. One person had a complete remission. While these results are still preliminary, when scientists are able to get the virus vehicle into more tumor cells, gene therapy such as this may become a widespread treatment for many different cancers. So there is a chance that Scully could be successfully treated by chemotherapy, radiation, and p53 gene therapy. Unfortunately, Dr. Scanlon is not the saint that he seems to be. Instead of using the proper treatment to aid the cancer-stricken women, he was hastening their deaths. Scully discovers his deception just in time. It is doubtful, therefore, that she received any gene therapy.

Few will argue that gene therapy doesn't belong in the arsenal of weapons to combat disease. The ethical debates on gene therapy, though, are only just beginning. One of the controversies surrounding gene therapy is whether to add genes in such a way that they will be passed on to future generations. Unlike current gene therapy, where only a few selected cells in a person are altered, every cell of a person will be changed if the new genes are introduced into eggs.

One of the issues concerning this type of gene therapy, called germ-line gene therapy, is that no one can predict where a newly added gene will land in the genome. The new gene could end up somewhere innocuous, or it could land in the middle of an important gene. The combination of adding a new gene and possibly disrupting some important gene in every cell of a baby whose opinion was not solicited could have consequences no one can foresee, like endowing a child with huge muscles, only to learn later that lack of strong bones to support the muscles cripples the child in early adulthood.

There are, however, circumstances where adding genes to eggs could have enormous advantages for a person and that person's descendants. One idea envisioned by John Cambell of the UCLA School of Medicine is to add a gene that causes a toxin to be produced only in prostate cells and only if a second substance is also present in the bloodstream. If a man develops prostate cancer, he could be given the second substance, which would trigger the toxin to be made in the prostate cancer cells, killing all the cells. It's fascinating to consider that added genes might someday eliminate cancers from humans by selectively killing off cancer cells if they arise.

INT. MICROBIOLOGY LAB—NIGHT

Scully is staring at the electron microscope talking on her cellular. Dr. Carpenter is sitting at the console where a new image now occupies the E.M. video monitor: A bacteria that has been sliced open exposing all its organelles. It is green in tint.

                    SCULLY
    It's me.

                    MULDER
    He's alive, Scully.

                    SCULLY
    Who?

                    MULDER
    The fugitive. The driver of the silver Sierra. He called the
    doctor's house when I was there.

                    SCULLY
    Where is he?

                    MULDER
    I don't know. Where are you?

                    SCULLY
    Georgetown microbiology department. I've got something for
    you.

                    MULDER
    Is it smaller than a silver Sierra?

                    SCULLY
    Much. And it's not silver. It's green.

                            —"The Erlenmeyer Flask"

Dr. Berubi's use of gene therapy in the episode "The Erlenmeyer Flask" is not as ethically defensible. As Scully tells Mulder in the episode, Berubi was probably growing bacteria in the Erlenmeyer flask to temporarily house virus that could be used as a vehicle in gene therapy experiments. As they are later to discover, the gene passengers for the virus vehicle are of alien origin. Gene therapy using alien DNA would be unlikely to receive government approval, a prerequisite for all current gene therapy protocols. Perhaps this is why Dr. Berubi was so secretive about his work. One can hardly imagine a review board at the National Institutes of Health approving a gene therapy permit for an experiment whose side effects include massive strength, ability to breathe under water, and thick green blood that contains toxic fumes.

When Dr. Scanlon is less than forthcoming about the origins of Scully's cancer, she decides to do some experiments on her own. In the episode "Redux," Scully and Dr. Vitagliano find evidence of a virus growing inside bizarre cells recovered from the arctic. Scully makes an intuitive leap that this virus, and not Epstein-Barr, is the true cause of her nasopharyngeal cancer. Chris Carter, who wrote "Redux," wanted Scully to conduct an experiment that would prove her cancer was connected with the virus in the strange cells. I told Chris that Scully could perform an experiment called a Southern blot, named after Dr. Frank Southern. This experiment allows Scully to determine if there is a match between her DNA and the virus DNA, which she would find only if the virus DNA had previously spliced itself into her genome. The basic procedure involves preparing samples of the virus DNA and her own DNA. After the virus DNA is made radioactive, Scully can determine if the radioactive virus "sticks" to her DNA. If it does, then a copy of the virus must already be in her genome.

After explaining the procedure in exhaustive detail to Chris, he asked me how long the experiment would take. I replied that if Scully was very efficient, maybe two days. Chris then told me that Scully had to complete the experiment in six hours. No way, I replied—not exactly the words that Chris wanted to hear. Still, he was adamant. She had six hours. So I conceded scientific accuracy to artistic license. However, I engaged in a last-ditch effort to save face among the 0.1 percent of the population who would know that this experiment could never be done in this amount of time. I suggested to Chris that Dr. Vitagliano tell Scully that he was skeptical that she could complete the experiment in the time allotted; at least not unless she could make the virus DNA, known as the "probe," extremely radioactive. In the episode, Dr. Vitagliano tells Scully that she will need a "blazing hot probe"—science slang for a very radioactive probe. After the episode aired, Chris's wife, Dori, told me that they were worried the line would not make it past the TV censors, something that I hadn't even remotely considered.[2]

In the summer of 1997, I discussed with Chris some cool new treatments for curing Scully's cancer. Scully's health was getting progressively worse, until she lay near death in the sequel to the episode "Redux," called "Redux II." Chris decided to have Scully "cured" when a synthetic chip is placed in the base of her neck. While watching the episode, I groaned. I knew what would happen the next day when I faced the five hundred rabid *X-Files* fans in my Introduction to Biology class. Hands flew up as soon as I entered the room. "How could a computer chip in the neck cure cancer, Dr. Simon?" I was asked repeatedly. When I talked to Chris later that week, he said, "It's not a cure! It's only remission!" I remain hopeful that some neat new medical technology will prevail in the end.

Spitting Images

According to FBI agent Fox Mulder, when cows are found mutilated and drained of blood, UFOs should top the list of suspects. Although Mulder is likely familiar with the exhaustively detailed police reports on such incidents, he is not impressed with their conclusions that such mutilations and blood loss can be explained by predators and scavengers. When Joel Simmons is discovered dead in Connecti-

[2]This experiment has apparently made Scully a legend in her own time. When I travel to universities to give seminars on my research, student scientists who know of my involvement with *The X-Files* invariably ask me for Scully's recipe for doing the Southern blot experiment so rapidly. Scientists in several laboratories have also told me that when an experiment is not working properly, their sympathetic lab mates tell them to "call Scully" since she can complete experiments in record time and with astonishingly clean results. With all the various concessions of scientific accuracy to dramatic storytelling in the series, it was amusing to discover that this is what the scientist *X-File* fans always point to as being highly inaccurate.

cut missing 75 percent of his blood, Mulder thinks that aliens have climbed up the tree of life to tackle more advanced specimens. Mulder is pleased to receive confirmation of his suspicions from Joel's eight-year-old daughter, Teena, who was nearby when her father was murdered. Teena tells a mesmerized Mulder that she saw red lightning and men from the clouds, classic validations of alien involvement. According to Teena, the aliens were out to exsanguinate her father—a curious choice of words from one so young.

The aliens must have been busy that day because across the country in San Francisco, another bloody drama was playing. The father of another eight-year-old child, Cindy, is also dead from exsanguination. Mulder and Scully are amazed to find that Cindy is the spitting image of Teena. The obvious interpretation—identical twins— is ruled out because the girls had different birth mothers. However, a connection between Teena and Cindy does exist: both mothers underwent in vitro fertilization treatments at the same clinic under the supervision of Dr. Sally Kendrick, who disappeared after being fired by the clinic for tampering with fertilized eggs before implantation.

Mulder is told by his mysterious government contact "Deep Throat" about a supersecret government project called the Litchfield experiment. It was, he explained, America's answer to the eugenics experiments by the Russians in the 1950s. The Russian breeding program, which matched athletes and scientists to produce supersoldiers, paled in comparison with the sophisticated experiments of the Americans—the cloning of boys named Adam and girls named Eve in a compound near the town of Litchfield. Mulder and Scully are sent to interview Eve 6, a long-term resident of a prison for the criminally insane. Eve 6, the mirror image of Sally Kendrick, tells Mulder and Scully that her extra strength and intelligence come from her five extra chromosomes—numbers 4, 5, 12, 16, and 22. This means, Eve explains, that she has fifty-six chromosomes instead of the normal human forty-six.[3] Eve omits mentioning which extra chromosomes

---

[3]If you're perplexed about Eve 6's math skills, how 46 plus the five new chromosomes equals 56, then you are probably in the same boat as most of the episode's audience. If, however, you remember from Chapter 3 that these chromosomes are all autosomes, and that autosomes always come in pairs of 2, then it should be clear that the Eves actually have 10 new chromosomes for a total of 56.

are responsible for her presence in a prison for the criminally insane.

According to the episode, the Eves were cloned in the 1950s. To understand how this might have been accomplished with the knowledge at the time requires a short digression into the history of cloning, which begins long before a sheep in Scotland became a worldwide celebrity. The quest to clone began in 1938 when Hans Spemann, the Nobel Prize–winning German embryologist, first proposed the experiment that was much later to be called cloning. Spemann's mind wasn't on producing genetically identical superhumans. Rather he and his contemporaries were intrigued with the question of whether the genetic material in a specialized cell still had what it takes to reconstruct an entire organism.

Spemann knew that the nucleus was the home of the genetic material, but whether genes were DNA or protein wasn't known until 1944. Spemann dreamed up the "fantastical" experiment of taking a nucleus out of a cell and transferring it into an egg cell that had lost its own nucleus. If an organism developed correctly from the genetic material provided by the transplanted nucleus, then the nucleus must contain all the information necessary to make the organism. If no organism developed, then the information in the transplanted nucleus must be deficient in some way.

Sadly for Spemann, the techniques needed to express his vision didn't emerge until after his death. In 1952, the technology existed for isolating tiny, fragile nuclei and for injecting the nuclei into egg cells without the nuclei or cells popping like pricked balloons. Embryologist Robert Briggs was the first to develop the techniques to isolate nuclei and transfer them into eggs. He chose to experiment on frogs because thousands of eggs can be easily obtained from a single female and the eggs are a thousand times larger than mammalian eggs. Briggs knew that a fertilized egg divides into two cells, then those two cells divide to give four cells, and so on until a ball of eight to sixteen thousand cells surround a hollow cavity. Cells in this stage of embryo development, called the blastula stage, do not appear physically different from one another. Since no obvious cell specialization has occurred at this point, Briggs thought that the DNA in these cells might still be capable of restarting the developmental process. To prove his hypothesis, Briggs replaced the nucleus in an unfertilized frog egg with a nucleus he had removed from a blastula cell. The egg with its new nucleus stated to divide and soon a small tadpole was swimming around. When Briggs used nuclei from a slightly later stage of development, after cells were thought to have begun specialization, development was abnormal and no tadpoles formed. Briggs's interpretation? Specialization leads to irreversible changes in the DNA, which makes it no longer capable of restarting the development program. Plausible, but wrong.

While Briggs's cloned tadpoles didn't swim into the evening news, the scientific world was astounded. Tadpoles could be cloned using nuclei other than ones that existed in fertilized eggs! With better techniques in the late 1950s, and a different species of frog, clones using blastula-cell nuclei were able to develop into adult frogs. Still, no matter who was doing the experiment, nuclei removed from cells of adult frogs led to tadpoles that croaked before metamorphosing into adults. Only nuclei from very young embryos could direct an egg to develop into a mature organism. Again, the most obvious interpretation was that the DNA in specialized adult cells was irreversibly changed.

Transferring nuclei into amphibian eggs was well established by the time of the "Eve" Litchfield experiments. However, the first report of transferring nuclei into mammalian eggs—those of rabbits—wasn't published until the mid 1970s. The technology of the 1950s was simply incapable of performing the delicate operation of placing a nucleus inside a tiny, fragile mammalian egg. If nuclear transfer was unavailable to the scientists of the Litchfield project, how could the Adams and Eves with their ten extra chromosomes have been cloned?

Let's start with how to possibly get extra chromosomes into an individual. The actual procedure isn't difficult. Spread the chromosomes from some human cell on a glass microscope slide, identify the chromosomes you want, and then carefully scrape those chromosomes off the slide. The chromosomes can then be injected into an unfertilized egg's nucleus. No big deal, according to a modest Jim Robl, who adds pieces of chromosomes in this fashion to mouse eggs. The tricky part for the Litchfield scientists would be keeping the chromosomes from breaking and also not damaging the egg with ten needle pricks.

Since this procedure requires little in the way of technology, it is not beyond the realm of possibility that the Litchfield scientists could have added extra chromosomes to an egg. Next, the scientists would need to produce cloned babies from the egg. Eve 6 shows Mulder and Scully the Eve family picture, eight little girls who look just like Teena and Cindy. So somehow, eight babies need to be made from that single egg.

There was a method to make multiple babies from a single egg that the Litchfield scientists could have performed in the 1950s. The scientists could fertilize their precious, extra-chromosome-containing egg with a sperm and let the egg start dividing in a culture dish. This part of the process is the same as that performed today in in vitro fertilization clinics throughout the world. While the first in-vitro-fertilized baby wasn't born until 1978, morals and ethics, and not lack of scientific acumen, kept such procedures from being performed sooner. Given the goals of the Litchfield experiments, the ethical hurdle for people involved in the project was less than daunting.

At the eight-cell stage, the cloning process would begin. To create eight identical children, the scientists would need to remove the stiff, jellylike protective coat that surrounds the eight-cell embryo by treating the little ball of cells with an acidic solution. Then, by gently agitating the sticky cells, they can be teased apart from one another. The eight single cells, all of them genetically identical, can each develop into an individual person. If embryos derived from the single cells were implanted into the uterus of eight surrogate mothers, eight identical babies with ten extra chromosomes would be born nine months later.

Actually, this last part is a bit of a stretch. Not the part where the embryo cells are teased apart and implanted individually in a surrogate mother (although having all eight cells survive the process would be incredibly lucky); methods of embryo splitting have been around for over fifty years and would have been available to the Litchfield scientists. Scientific accuracy breaks down by having humans survive with extra autosomal chromosomes. As described in Chapter 3, extra autosomal chromosomes are poorly accepted by a developing fetus, whether added by genetic engineering or the consequence of a quirk of nature. The only additional autosomal chromosome that a human fetus can survive with is a single extra copy of chromosome 21. This is known as Trisomy-21, for the three copies of chromosome 21. It is also known as Down's syndrome. Two extra copies of chromosomes 4, 5, 12, 16, and 22 would cause irreparable harm to the fetus. Also, Eve 6 implies that the extra chromosomes provided her with heightened strength and intelligence. The scientists who created the Eves would have needed to know which genes contributed to strength and intelligence and which chromosomes contained these genes. In the 1950s, no genes had been mapped to any autosomal chromosome and even today, no one understands all the genes that contribute to complicated traits like strength and intelligence. But this, of course, is what puts the fiction into science fiction.

Seeing the similarity between the young Eves and Teena and Cindy, Mulder realizes that Sally Kendrick was indeed manipulating eggs at the in vitro fertilization clinic. She was cloning herself. To make the clones, Sally needed eggs, and what better place to find eggs than an in vitro fertilization clinic? Sally must have used nuclei from her own cells and transplanted them into the eggs of Teena's and Cindy's mothers. But Sally was trying to do more than simply clone herself—she was trying to correct the defects in the Eves, the psychoses that began at the age of sixteen and turned them homicidal at the age of twenty.

To correct the defects, Sally would need to make specific changes to the Eve genome. Sally must have been interested in determining if the increased strength and intelligence were possible without the mental aberrations. Curing her own problems using gene therapy wasn't an option since her difficult-to-reach brain cells would need fixing. What Sally could do is change the genome of one of her cells, and then clone children from the nucleus of that cell. Sally had good intentions when she tried to correct the mental abnormalities present in the Eves, but learns too late the uncertainties that may be associated with germ-line gene therapy. What she succeeded in doing was creating eight-year-old full-fledged homicidal monsters, little girls who eliminate people like their fathers, Sally, and almost Mulder and Scully for no better reason than they want to.

It is now clear that nature permits the cloning of mammals and the technology already exists to clone sheep, cows, and mice. Most scientists agree that if sheep, cows, and mice can be cloned, so can humans. One question is whether such a feat might have already been achieved. In 1978, the book *In His Image: The Cloning of a Man* was published by J. P. Lippincott. This very successful book tells the story of Max, the aging millionaire who wishes to have a clone as his heir. In 1974, Max convinces a noted science writer to find him a scientist who can give him the heir that he wants. A lab is set up in a Third World country. Women in the vicinity donate eggs and the millionaire supplies some of his cells. The heir is born two years later. What differed about this book and the hordes of other cloning sagas was that the author, David Rorvik, claimed the book was nonfiction, and that a clone really was born in 1976.

The book describes in accurate detail the science behind cloning procedures, at least what was known in the late 1970s. The details of the procedure beyond what was known are hazy, leading naturally to widespread skepticism in the scientific community. Rorvik, a former science reporter for *Time* magazine, never saw the putative baby clone, nor was the science of the time advanced enough to prove genetically that any baby born was a clone. Rorvik continues to maintain that the book was not fiction, most recently in an interview with *Omni* magazine. Walking around somewhere, he insists, is a twenty-something clone, the spitting image of his father . . . and maybe there is a mental institution somewhere occupied by six women named Eve and children named Teena and Cindy who are also the spitting images of one another.

All the Buzz on Bees, Corn, and Viruses

INT. HOSPITAL MORGUE

                    SKINNER
Smallpox?

                    MULDER
According to the coroner, an especially virulent strain,
caused by a mutant variola virus—

                    SKINNER
Caused how? How could this man have contracted a disease
that doesn't exist anymore?

Mulder hands him a small, clear container that sits on an adjacent table.

MULDER
From these.

Skinner examines the contents—what appear to be several dozen tiny fibers.

SKINNER
What are they?

MULDER
Bee stingers and venom sacs. Recovered subcutaneously from
the victim's face, neck, and arms—

SKINNER
You're saying this man was stung by bees carrying smallpox?

—"Zero-Sum"

If you saw the *X-Files* movie and are unfamiliar with the *X-Files* series—or even if you are familiar with the series—you may be puzzled about how thousands of bees, acres of corn, and deadly viruses are connected. The explanation requires revisiting the fourth-season episode "Herrenvolk," which begins with an unfriendly encounter between a bee and a telephone repairman. The repairman, fixing a faulty line in rural Canada, is mildly annoyed when the bee stings him on the neck without cause. Looking down from his perch high atop a telephone pole, he is surprised to find five boys, identical in both their silence and their appearance, watching him. The repairman is suddenly shaken by seizures and retching, causing him to fall to the ground, where he dies as the expressionless children walk away.

Mulder is taken to the location where the repairman died by Jeremiah Smith, a mysterious other-than-human individual who works for the Social Security Agency. The repairman's withered body is covered with sores and scabs, even though the man was alive only twenty-four hours earlier. Jeremiah leads Mulder past the body into extensive fields of flowering plants that are being grown, he explains, for their pollen. Mulder is shaken to his very core to see his sister, Samantha, working in the fields, yet looking like the child she was when she was kidnapped twenty years earlier. The strangely silent and expressionless Samantha is accompanied by one of the boys who witnessed the death of the repairman. Mulder becomes even more confused when he sees that all the children in the area are identical to either Samantha or the boy. Jeremiah Smith explains to Mulder that they are clones, drone clones, whose duties are to tend the fields and a large colony of bees that inhabit an expansive apiary.

It is an interesting twist, human drones taking care of bees, some of which are also drones. Of the twenty thousand bees in a typical beehive, about two hundred are males who develop from the unfertilized eggs of the queen, their mother. These males, known as drones, have only half as much DNA as their sisters, who are generated from fertilized eggs. Development of organisms from unfertilized eggs, called parthenogenesis, was discovered over 250 years ago by a man named Bonnet, who must have spent many fascinating hours determining the sex of thousands of aphids before he concluded that they, like bees, can produce young from eggs that are never fertilized.

The bees that would have stung Mulder in "Herrenvolk" had he not doused himself with gasoline are all females; insect stingers are actually converted oviducts, an organ normally used to lay eggs. Stinging to protect the hive is just one of the many jobs of female worker bees. They are also responsible for finding the food, doing the housework, caring for larvae, air-conditioning the hive, and hand-feeding tidbits to young males and the queen mother. With all this activity going on, female workers last only about a month before dropping dead from exhaustion. The male drones hang around the hive getting fat while waiting for a chance to have sex with an unrelated virgin queen and die. Although difficult to understand in human terms (although possibly not by male readers), this social structure must have something going for it because honeybees have been acting this way for over 20 million years.

Bees and aphids still engage in sexual as well as asexual reproduction. Imagine, though, if an organism reproduced only by cloning. Every member of the species would be female; every descendant identical to its mother. The eggs of the mother would require no fertilization since they already contain a whole genome—the exact genome of the mother. While this may sound like the plot for a B science fiction movie like *Amazon Clones from Venus*, it isn't. It's a description of whiptail lizards of New Mexico. The lizards seem perfectly normal until you realize that they are all female and the babies are all identical to the mothers.

Cloning is the natural way that a number of fish, amphibians, and reptiles produce baby fish, amphibians, and reptiles. Reproduction by cloning is widespread in the animal kingdom probably because exceptionally fit females can be continuously generated without diluting their genome with that of a weaker male. Also, by not needing two to tango, time and energy aren't spent producing males or on courtship and mating, allowing females to range long distances looking for a nice place to raise their genetically identical babies. There is also little chance that an excellent genome will be lost due to the death of an animal before it can reproduce—countless siblings are still around to pass along the family genes. In fact, there seem to be so many advantages to reproduction by cloning that entire books have been written to answer the question of why more organisms don't reproduce exclusively by cloning.[4]

Part of the answer to why more organisms don't reproduce by cloning is that cloning permits little genetic diversity—there is no mixing of male and female genetic material to produce offspring with unique new genomes. Species that reproduce by cloning therefore tend to be very similar genetically. The clones are probably all related to the one female that acquired the original mutation allowing it to produce eggs which don't require sperm to start development. Without the constant genetic mixing that is the hallmark of sexual reproduction, only the occasional mutation in the DNA of an egg allows for the further evolution of the species. The genetic uniformity of a clonal species is disastrous if a disease strikes the population and no members have genes that allow the species to survive. This same lack of genetic diversity in artificially created clonal populations disturbs some people confronted with the new era of animal cloning. While Bessie the cow may have many wonderful characteristics, she may

[4]For example, Roger N. Hughes, *A Functional Biology of Clonal Animals*, Chapman and Hall, New York, 1989.

also be susceptible to cow diseases that might not be as deadly if af-flicting an unrelated cow. If all cows were Bessie clones, cows as a species could be wiped out by a single nasty bacterium or virus.

Mulder learns that the repairman died from a deadly bee sting. However, before he can determine what made the bee lethal or how the bees on the Canadian farm are connected with the plants being grown by the cloned children, he is forced to run for his life from a bounty hunter who wants to kill Jeremiah Smith. When Mulder is fi-nally able to return to Canada, the fields and apiary are gone.

The bees resurface in the episode "Zero Sum" later in the fourth season. "Zero Sum" begins with a postal worker in Virginia sneaking a cigarette in an otherwise unoccupied ladies' room. The lone smoker is soon joined by thousands of bees swarming out of the walls and fixtures, apparently alerted by guard bees that someone is smoking in a nonsmoking area. The woman, who doesn't fully appreciate the dangers of smoking, is startled by the bees and then does exactly what experts say not to do: she stands still and whacks at them in-stead of running away. Not surprisingly, this further upsets the bees, which respond in typical bee fashion by stinging their attacker re-peatedly. The bees then disappear back into the walls.

The quickness with which the bees emerged from the walls and the sheer numbers that left the nest suggest that these bees are Africanized honeybees, the so-called killer bees. Africanized honey-bees were imported to Brazil in 1956 and escaped from disenchanted beekeepers a year later. On the move ever since, the bees travel about two hundred miles a year, attacking and killing European hon-eybee queens and taking over their hives. Africanized honeybees and the slightly larger, more docile European honeybees are very similar in appearance, have exactly the same venom, and comprise a single species. Although Africanized honeybees are highly aggressive, easily disturbed, and like to sting repeatedly, they have caused far less may-hem than depicted in movies like *The Savage Bees* (1976), *The Killer Bees* (1974), and *The Swarm* (1978). The bees have, however, killed six elderly people and a number of pets since reaching the United States in 1990.

Since an Arizona electrician survived being stung seven hun-dred times, the postal woman in "Zero Sum" may not have received a lethal dose of venom. A detective viewing the strange marks on the body is convinced that the woman died from something other than bee stings. Confronted with this mystery, he e-mails pictures of the body to a certain FBI agent who deals in unusual cases. Before Mulder can investigate, the pictures are erased from his computer by his boss, Assistant FBI Director Walter Skinner, who travels to Virginia, incinerates the body, and covers up the crime scene. Although Skinner is acting reluctantly in the hope of gaining a cure for Scully's cancer from the malicious Cigarette Smoking Man, he cannot help but do some investigating on his own—especially after being framed for the murder of the detective. Skinner is surprised to find that the walls of the mailroom rest room are impregnated with bee combs. Tearing off a chunk, he brings it to a forensic entomologist, who coincidentally was also consulted by Mulder after Mulder returned from the Canadian farm.

It actually isn't surprising that Skinner and Mulder contact the same forensic entomologist considering how few of them there are. Entomology is the study of insects, and forensic entomologists are scientists who find connections between insects and crimes. Given the paltry number of crimes actually committed by insects, forensic entomologists also study the insects that arrive subsequent to the crime. By examining these insects, a forensic entomologist can pinpoint the time and sometimes the cause of death and possibly even help catch more human killers.

When investigating a crime scene, forensic entomologists look for metallic-green blowflies, since these insects are usually the first to arrive after a person dies. Insects such as blowflies smack their lips when they see a dead body. If the blowflies have laid eggs in the body, then the body must have been dead for at least two days. The time of death can be specified even further by determining how old any insect larvae are. No insects present? The body was probably frozen or wrapped. Some insects lay their eggs only indoors; others only outdoors. Finding the wrong kind of insect on the body usually means that the body was moved. If the body isn't found for a while, there may not be enough tissue left to run toxicology tests. But the same tests can be run on the insects who have taken up residence since any drugs in the body will now be in them ("The maggots were high on heroin, Your Honor!").

The forensic entomologist helping Skinner determines that the chunk of bee comb isn't empty when he pulls out a larval bee with his forceps. And not just any larval bee, but one coated in royal jelly—an immature queen bee. The entomologist uses the young queen to generate more bees for identification purposes. Soon, the entomologist makes the supreme sacrifice asked of so many *X-Files* scientists as he is killed by hundreds of bees who aren't happy being cooped up in a small container. This time the body doesn't disappear before Mulder gets a good look, and he is convinced that the pustules and marks on the entomologist, the same as in the photographs of the postal woman, are not just due to bee venom. Mulder tells Skinner that the entomologist had symptoms of smallpox, a disease caused by the variola virus.

Mulder is correct when he tells Skinner that the variola virus has killed more people than any other disease agent in history. The 7 million people who died of smallpox in A.D. 180 coincided with the decline of the Roman Empire. One hundred years after the Spanish introduced the virus to Mexico, smallpox brought down the Inca and Aztec empires, leaving only 1.6 million people alive from the original population of 25 million. Four hundred thousand people a year— kings and queens, princes and peasants—died of smallpox in Europe up to the late eighteenth century.

Compared with my tiny turnip crinkle virus with its five genes, variola is a huge, brick-shaped monster of a virus with two hundred genes in its DNA genome. The virus is so immense that it can almost be mistaken for a strange-looking bacterium using a light micro-scope. The variola virus is the most adept of all human viruses—a master at infecting people and evading the immune system. The well known symptoms of the virus would appear suddenly. Fever and limb pain were followed a few days later by a rash that covered the face and eventually the entire body. The rash turned into skin lesions and then into pustules. Lucky survivors were rewarded with disfig-urement and frequently blindness.

The telephone repairman in "Herrenvolk" and the postal worker and forensic entomologist in "Zero Sum" died within minutes of being stung by the bees, their faces covered with pustules. Mulder speculates that the bees somehow were responsible for transmitting the virus. Since smallpox symptoms normally take ten to twelve days to develop, Mulder believes that they were infected with a mutant form of the variola virus, a virus that is now deadlier than ever. While Mulder may be correct, it isn't clear how such mutations got into the virus. Using genetic engineering methods to create mutations in specific viral genes to speed up the action of the variola virus is beyond the knowledge of earthly virologists. Making mutations in viral genes is not the problem. Creating mutations in a particular gene of an organism is a routine method of scientific investigation used to learn about the function of different genes. However, without knowing which of the variola virus's two hundred genes helps the virus to evade the immune system and produce the symptoms that it does, it would be extremely difficult to rationally design mutations that could speed up the virus's ability to invade a person.

It is also unlikely that this particular virus could naturally mutate on its own, at least now. Mutations imply a change in the order of nucleotides in the DNA of the virus—something that can occur naturally only when the virus is reproducing inside human cells. Yet one of the great achievements in medical history is the complete eradication of the variola virus from nature, which eliminates any opportunity for the virus to mutate by itself.

The conquest of smallpox is mankind's first triumph over a disease. Long before a vaccine for the variola virus was developed, it was common knowledge that people who recovered from smallpox were forever free from any reoccurrences of the disease. The Chinese were possibly the first to infect people with a second, much more benign strain of the variola virus to protect them against the deadly strain. By the early eighteenth century, this process, known as variolation, had spread to Europe and became popular when crowned heads of state variolated themselves, their families, and their soldiers. Variolation, however, was not without risk: about 2 to 3 percent of people who were variolated died of smallpox from the less deadly strain. Still, this was a vast improvement over a death rate as high as 90 percent for children if they became infected by the more virulent strain of the virus.

Variola is one member of a family of viruses called poxviruses. There are poxviruses that infect birds, sheep, pigs, monkeys, camels, rabbits, buffalo, and cows. Although variola virus only infects humans, some of the animal poxviruses can weakly infect humans as well as their main animal host. All outer coverings of poxviruses—

the "bag" that surrounds the virus genome—are very similar. This means that people who survive infection with one poxvirus have their immune systems primed to destroy any further infecting poxviruses.

One of the most common poxviruses is cowpox, unsurprisingly named for its primary target. If there was a benefit to being a milk-maid in eighteenth-century England, it was that many became infected with cowpox virus after handling diseased udders, protecting the young ladies from ruining their looks or their health by later coming down with smallpox. Knowing this, a farmer and a school-teacher used cowpox virus to immunize relatives and family friends. Unfortunately, slicing open arms and rubbing the wounds with pustules from a dirty cow led to some unpleasant infections and a consequent reduction in the number of further volunteers. In May of 1796, Edward Jenner, a doctor, used more sanitary methods to test the efficacy of immunization with cowpox virus. He was able to demonstrate scientifically that treatment of people with fluid from a cowpox pustule prior to variolation kept the normally mild variolation symptoms from developing. Jenner had scientifically proven the efficacy of vaccination—that deliberately infecting a person with a weak or mild strain of a virus can prime the immune system to protect the person against more deadly strains of the same or similar viruses.

Jenner was able to convince a skeptical medical community that he had found a safer way of protecting people against smallpox. The news of the vaccine spread quickly. By the beginning of the nineteenth century, 100,000 people had been vaccinated across the globe. In 1806, Thomas Jefferson correctly predicted that smallpox would some day be remembered only as a historical artifact. Yet as late as the mid-1960s, smallpox was still killing 2 million people a year, a tragedy made monumental since the vaccine was so effective.

In 1967, the World Health Organization developed the goal of completely eradicating smallpox from the planet. Highly stable freeze-dried vaccine, a new inoculation device and a strategy change from mass vaccinations to surveillance and containment vaccinations, succeeded in eliminating smallpox in only ten years. The last person to catch smallpox from contact with an infected person was a cook in a Somali village in 1977 (he survived). The dubious distinction of being the last person infected with the variola virus was a British photographer, who died of smallpox after being exposed in a laboratory mishap in 1978. On May 8, 1980, the World Health Organization declared that smallpox as an illness had been wiped off the face of the Earth.

The variola virus's mastery of human infection was also its downfall. Once people discovered how to protect themselves against the virus, it had no place to hide. Unlike other viruses, such as the influenza (flu) virus, which also infects pigs and birds, variola only infects people. When the last person to become infected couldn't transmit the virus because he was surrounded by immune medical workers, the virus died out everywhere except in laboratories. The last samples of the virus are currently hidden away inside high-security facilities at the Centers for Disease Control in Atlanta and at the Institute for Viral Preparations in Moscow.

While Mulder was making the first connections between bees and the variola virus in "Herrenvolk," Scully was finding a disturbing link between smallpox vaccinations and a large number of unexplained data entries in Social Security Agency computers. Along with FBI agent Pendrell, Scully discovers that computers used by Jeremiah Smith and several missing Social Security employees contain a billion file entries that all begin with the initials SEP. Scully learns from one of Mulder's government contacts that the letters stand for Smallpox Eradication Program.

The Smallpox Eradication Program mentioned in "Herrenvolk" is real. In 1981, the World Heath Organization recommended destroying the few remaining stocks of the variola virus. Three reprieves later, the new date for the destruction of the virus is June 30, 1999. Destroying the last of a virus that has caused such misery throughout history seems like a no-brainer. Eliminating the virus would erase forever the fear that terrorists could use variola as a biological warfare weapon. Since children are no longer vaccinated against smallpox, the death rate from such an attack would be horrific.

But some scientists think that destroying the last stocks of the virus is a mistake; that once the virus is gone it is gone forever. The variola virus would be the first species on our planet ever to be deliberately destroyed. Some very prominent virologists are concerned because the variola virus can still teach us about how viruses are able to circumvent the immune system—information that could be important in combating other deadly viruses like HIV and Ebola. Since the virus is safely stored in Atlanta and Moscow, why destroy it if one day it might prove useful? This is the perfect question for someone whose mind is always searching for interesting government "conspiracies." Chris Carter must have thought that the Smallpox Eradication Program was a natural for a subplot in *The X-Files.*

After seeing the computer entries with the designation "SEP" followed by letters of the alphabet—twenty different letters—Scully makes a brilliant leap of logic. As a trained scientist, Scully knows that proteins are chains of twenty different amino acids, each represented by a letter of the alphabet. For example, the amino acid alanine is known as "A"; the amino acid methionine is "M," and the amino acid glutamine is "Q" (no one ever said that scientists could spell). When Scully sees the twenty-letter code, she would undoubtedly think "protein amino acid sequence"—at least, that's what I told Chris Carter she would think.

Once Scully realizes that the letters following SEP are probably the order of amino acids—the sequence—of a particular protein, she uses the Internet to search for the identity of the protein. The sequences of proteins that have been analyzed so far are stored in huge databases that can easily be accessed by anyone with a modem. Scully must have entered the series of letters into her computer and then used the Internet to search the database for a match . . . which she found. The order of the letters following the SEP designation corresponded with the amino acid sequence for one particular protein, a protein that makes up the "bag" that surrounds the genome of cowpox virus.

Scully would know that cowpox was the original virus used to inoculate people against smallpox. She then makes another intuitive leap. Somehow the data entries are related to people's smallpox vaccinations. Scully, like everyone born before 1980, would have the telltale mark of being vaccinated against smallpox. She examines her smallpox vaccination scar and that of Agent Pendrell and determines that the scars contain remnants of the vaccinated virus in a nonrandom pattern. In other words, the vaccination was much more than simply the injection of a mild virus to protect against variola. It was part of a scheme to affix a tag to her and everyone else in the world through their vaccinations.

Chris Carter and I had several long discussions about the science in this episode. Chris called me in June 1996 with the following question: How can you "tag" someone with their smallpox vaccination? This was a tough question because, frankly, you can't tag anyone with a vaccination. Since this was not the answer that Chris wanted, I suggested implanting a computer chip. Already done that, Chris replied. Since I couldn't come up with anything else off the top of my head, I said that I needed time to think about it. Chris said that he would call back tomorrow and, oh, by the way, it had to involve computers.

So I spent the next several hours walking my dogs in circles around a Little League baseball field trying to come up with the least preposterous way to tag people with their smallpox vaccinations. The best that I could come up with was the following: aliens—really, really smart aliens—devised a method whereby the vaccine virus could be injected into the arms of people in a precise, predetermined pattern much like using a stamp device and ink to impress a bar code on a piece of paper. The "bar code" imparted by the injected virus, somehow still present after all these years, cannot be seen with the naked eye for the virus "ink" is much too small. However, the location of the virus can be visualized by using a very real scientific technique with the tongue-twister name of immunohistochemical staining.

Understanding the principle behind immunohistochemical staining is much easier than pronouncing it. Let's say that there are small scraps of paper on a football field between the forty- and fifty-yard lines. If you want someone sitting in the top row of the bleachers to see the paper, you are going to need to make the paper more visible. So you make a set of black clips that seek out and attach themselves to every piece of paper. The black clips are like antibodies, antibodies that specifically recognize paper and then attach themselves to the paper. So wherever there is a black clip, there is also a piece of paper. The black clips, however, are still too small to be seen from the top of the stadium. So you make a second set of clips—blue clips—that seek out and attach themselves to the black clips. The blue clips are antibodies that specifically recognize and stick to black clips. Before the blue clips attach themselves to the black clips, you tie a large red balloon to each blue clip. Now, wherever a blue clip attaches to a black clip, a red balloon marks the spot. So wherever there is a red balloon, there must be a blue clip attached to a black clip attached to a piece of paper. Now anyone sitting high up in the bleachers would know that the paper is between the forty- and fifty-yard lines because that's where all the red balloons are.

This is the idea behind immunohistochemical staining. Scully treats her vaccination scar slice with one kind of antibody (the black clips) that recognizes and attaches to a particular protein on the virus that was used for the vaccination. Then Scully uses a second antibody (the blue clips) that recognizes and attaches to the first antibody. Linked to the second antibody is a chemical that produces a bright color (the red balloons) which is easily visible using a microscope. So everywhere Scully "sees" the color in her vaccination scar slice, the virus must also be present. Scully uses a confocal microscope to visualize where the color is, since this type of light microscope produces a beautiful, colorful three-dimensional image of the vaccination scar slice. Scully finds that the location of the virus in her vaccination scar is not the random pattern that it should be but is very similar to Agent Pendrell's smallpox vaccine virus pattern. My idea was that differences between the virus patterns in Scully and Pendrell's vaccination scars account for the physical differences between her and Pendrell: she is a woman, red-haired, etc. and was vaccinated in a different town. But because they are both Americans and Caucasians, the patterns would be similar as well.

After explaining to Chris how Scully could use immunohisto-chemical staining to find that her vaccination was a type of identification tag, he said finc, but what about the computer? I explained that the entries in the computer could be the order of amino acids in a protein—one of the proteins of the virus that was used for the smallpox inoculations. The idea was that the computer entries would be slightly different for different people, meaning that the virus used to inoculate people against smallpox was slightly different for each person.[5]

---

[5] I wouldn't spend a lot of time trying to make sense of this intricate (and impossible) way of tagging people. The virus used for Scully's vaccination would be long gone and therefore Scully would find no "bar code" in her vaccination scar. The scar is merely the ghost of symptoms long past, a souvenir from being vaccinated with a very mild poxvirus. It was, however, the best "tag" I could come up with at the time given the parameters of smallpox vaccines and computers.

It was my idea to use the real amino acid sequence of a cowpox virus protein (the letters following the initials SEP) when it flashes on Agent Pendrell's computer screen in the episode. This was purely for my own scientific satisfaction and I was very pleased when Chris went along. Why this anal attention to scientific accuracy, especially in light of a rather preposterous way of tagging people? Because I knew that scientists are big *X-Files* fans, and I knew that some would freeze their television screens on the protein sequence that appeared on Pendrell's computer screen, copy down the sequence, and use the Internet to determine if it was a real protein sequence or just some made-up string of letters.

The reason for my professional paranoia comes from a well-known (in scientific circles) incident surrounding Michael Crichton's book *Jurassic Park*. In the book, Crichton presents the DNA sequence (order of nucleotides) of a gene that is supposed to be a dinosaur gene. Naturally, many scientists wrote down the sequence and searched through a DNA sequence database to see if it was related to any known genes. If Crichton had used the sequence of an authentic reptile gene, maybe with a few changes to simulate an extinct "dinosaur" gene, his reputation among scientists would have soared to heights comparable to that of the film version's box office gross. But sadly, Crichton used a well-known DNA sequence from a common laboratory bacterium. My hope was that some of the people who were disappointed with the dinosaur DNA sequence in *Jurassic Park* would search the protein database with our amino acid sequence and be impressed to find that it really was a cowpox virus protein.

Think this wouldn't happen? Guess again. A few weeks after "Herrenvolk" aired, a friend who teaches at Indiana University (and who did not know about my connection with the *X-Files*) was giving a guest seminar at my university. During dinner that night, he told me about how he had used an *X-Files* episode to teach his class about immunohistochemical staining. He said that he also decided to show the class how the amino acid sequence displayed on Pendrell's computer screen couldn't possibly be the sequence of the protein that it was supposed to be—having little faith that the show would go through the effort of using a real sequence. My friend demonstrated to his students how to enter the amino acid sequence into the computer and search the protein sequence database. To his astonishment, the protein sequence was exactly what the episode said that it was. My friend said both he and his class were speechless.

Back to the bees and viruses. Mulder has the bee stingers and venom sacs analyzed in "Zero Sum" and finds that they contain traces of the variola virus. The bees, therefore, are transmitting the virus as they sting people, who then succumb to smallpox and not bee venom.

Skinner learns that the bees are in the Virginia post office by mistake. Their true destination was elsewhere, but the package was damaged in transit. The Virginia post office was the way station for damaged packages, which end up in a storage room next to the ladies' rest room. Mulder's contact at the United Nations tells Skinner that seven packages were shipped from Canada to South Carolina, but one ended up damaged in Virginia. What became of the other six packages is soon made clear. The bees are deliberately released into an elementary school playground in South Carolina, where they wreak havoc with children and their teacher. Skinner tries to convince the doctors at the hospital emergency room to treat for smallpox but the doctors ignore him because, after all, smallpox doesn't exist anymore and the children became sick much too fast.

Skinner realizes that the bees are being used as carriers of the virus. How the bees picked up the virus is one of those instances where scientific accuracy gives way to good dramatic storytelling. Bees simply do not transmit diseases beyond an occasional fungus picked up from one flower and carried to another. However, it is conceivable that bee stingers could contain traces of the virus—especially if the virus was present inside pollen that the bees picked up.

Chris wanted the fields of flowering plants in Canada to produce the variola virus in their pollen; so he called and asked, is it possible to produce virus in pollen? The answer to the question is yes. If the virus genome was spliced into the genome of the cells of the plant and a mechanism was added to pop the virus out of the genome in pollen, then a self-contained virus like variola might replicate inside pollen and produce a virus that could be infectious. Adding an enormous virus like variola to the genome of a plant cell wouldn't be easy, but it could theoretically be done. Plants, just like animals, can be genetically engineered to contain additional genes—and a virus is, after all, just a bag of genes. Plants are actually much easier to genetically engineer than animals. If you want a new gene in every cell of a plant, then you simply add the gene to the genome of one cell and then grow a plant from that one cell. Recall that to generate an animal with a new gene present in every cell, you need to add the new gene to the genome of a cell that knows the make-an-animal program. This cell needs to be an egg or a cell not too many divisions away from the original egg. In plants, just about every cell of a mature plant knows how to make a complete plant. This means that you can add genes to the genome of a single cell, like a leaf cell, and then grow a plant from that cell.

Adding genes to the egg of an animal requires complex equipment and the success rate is not high. Adding genes to a plant cell is trivial—thanks to the bacterium *Agrobacterium tumefaciens*.

When it was originally discovered that *Agrobacterium* transfers some of its own DNA into a plant cell, a clever idea was born. Why not use *Agrobacterium* as a vehicle to transfer other DNA into a plant cell? The technique worked wonderfully. Splice one or more genes, or even an entire virus genome, into the DNA of *Agrobacterium* and when the bacterium transfers its own DNA into plant cells, the DNA that you added is transferred as well. The bacterial DNA and any DNA attached to it goes into the nucleus of the plant cell, where it splices itself into the genome of the cell. When a plant is grown from that cell, it will contain the new DNA in all its cells.

The ability to genetically engineer plants, to make so-called transgenic plants, has caused an entire new industry to spring up, known as agricultural biotechnology. Thousands of acres of farmland are already covered with cotton containing precolored fibers, tomatoes that don't soften as they ripen, potatoes that resist insect destruction, and soybeans that can survive herbicide treatments. One of the most exciting uses of the new technology is the creation of plants that produce pharmaceuticals at a fraction of the cost of producing the same products in the laboratory or in genetically engineered animals. These plants become little factories, producing products like antibodies ("plantibodies") to fight cancer, and proteins to use as vaccines. Someday in the not-too-distant future, vaccines administered by expensive, unpleasant injections will be stories to frighten your grandchildren with as you simply eat a banana or a baked potato to become immunized against cholera, the *E. coli* bacteria, or viruses. For countries where refrigeration and delivery of vaccines is a problem, edible vaccines may save the next generation of children from death or disfigurement due to preventable diseases.

In "Zero Sum," the deliberate release of virus-carrying bees in a South Carolina school yard is a test by a sinister group of human conspirators that are allied with the aliens that Mulder has spent his adult life searching for. The conspirators are testing the use of bees as a delivery system for viruses. But why a mutant variola virus? Although the series has not yet provided a complete explanation, killing people with smallpox is not the grand scheme. The virus that the conspirators have in mind for the bees to eventually disseminate is the virus present in the black slime unearthed in Tunguska and in an underground cave in Texas. As discussed in Chapter 3, this virus has a habit of producing alien entities with pointed teeth from the cells of the human chest cavity.

Which brings us to the corn. Midway through the *X-Files* movie, Mulder and Scully come across acres of corn and a large apiary. Before they can ponder any possible connections between bees and corn, they are chased from the apiary by the bees and then through the cornfields by black helicopters. Mulder finally learns of the relationship between the bees and the corn from an old colleague of his father, a man named Kurtzweil. Kurtzweil tells Mulder of a plot to unleash a terrible plague upon the people of this planet. Mulder is shocked to learn that the conspirators plan to infect people everywhere with a virus by means of the bees. The corn is for production of the virus in pollen that the bees will pick up and disseminate by stinging people. And the only survivors will be the aliens, alien-human hybrids, and the human conspirators and their families, who have been exposed to a new vaccine.

People who are acquainted with bees will most likely think that corn is a rather odd choice for a plant whose pollen needs to be picked up by bees. Corn is pollinated by wind, not bees, who prefer colorful flowers. Corn was chosen by Chris for its imagery; can you imagine the drama of Mulder and Scully being chased through fields of rosebushes or daisies? And it's not as much of a stretch as all that; although bees don't pollinate corn, bees find corn pollen to be very tasty and have therefore been found with corn pollen on their bodies.

Corn and similar plants are difficult to genetically engineer using *Agrobacterium;* the bacteria simply prefer to infect other types of plants. Since it's difficult to badger the bacteria into infecting plants like corn, a piece of equipment known as the "gene gun" was invented. DNA that you want to insert into a plant is applied to the surface of tiny tungsten bullets that are shot out of the gun and into plant cells. Inside the cells, the DNA is released from the bullets and travels to the nucleus and inserts into the genome. Cells that receive the bullets and their DNA coating are then used to grow a new plant. That plant will produce seeds that also contain the new DNA. If the DNA that coats the bullets is virus DNA and the bullets are shot into corn cells, whole fields of corn containing the genome of the virus can be produced. Add the correct DNA "switches" to the virus genome, and the virus can be made to pop out of the genome of pollen cells. In a science fiction setting, the pollen and its virus inhabitant can then be picked up by bees and transmitted to people through stinging. As Kurtzweil explains to Mulder in the *X-Files* movie, the corn is for production, the virus is the product, and the bees are for transportation. A deadly sting that the conspirators thought would depopulate the world will instead repopulate it with virus-induced alien life-forms.

And that's all the buzz on bees, corn, and viruses.

5

Seeking the Fountain of Youth

Introduction

INT. HALLWAY OUTSIDE DR. AUSTIN'S OFFICE—NIH

Scully and Mulder exit into the busy hallway.

> SCULLY
> You don't reverse aging, Mulder.

> MULDER
> Ridley's found a way.

> SCULLY
> Listen to what you're saying.

> MULDER
> He wanted human research subjects, right? Prisoners. Prisoners like John Barnett.

> SCULLY
> Mulder... It's science fiction.

MULDER

Twenty years ago, what would you have said about gene splic-
ing? Cloning? DNA fingerprinting? Artificial intelligence?
(Scully can't deny it.)
Maybe we're not looking for a man in his late forties. Maybe
John Barnett's found the perfect disguise. Youth.

—"Young at Heart"

In 1908, French scientist Alexis Carrel began an experiment that
was to stun the world. Working at what is now Rockefeller University
in New York City, Carrel had a simple goal: he wanted to know if tissue removed from an organism could be kept alive in the laboratory
longer than the normal life span of the organism that it came from. To
answer the question, Carrel carefully sliced off a tiny blob from the
heart of a chicken and glued it to the base of a sterile glass container.
After he added a nutritious broth, the cells at the fringes of the heart
fragment began to divide . . . and divide . . . and divide. When the tissue expanded to fill the bottom of the glass container, Carrel transferred a small amount to a new container where the cells continued to
divide. They kept dividing while the *Titanic* was sinking. They divided
through World War I. Through the stock market crash of 1929 and the
years of the Great Depression. Only the death of Carrel in 1944 caused
his assistant to terminate the remaining cells. If all the tissue that was
derived from the original chicken heart had been saved throughout
the thirty-four-year experiment, rather than being regularly discarded, it would have needed a container 800,000 miles across.

Since it's a rare chicken that reaches the ripe old age of twelve
years, Carrel had discovered that cells can long outlive the organism
that they came from. And if cells removed from an animal never died,
that meant some detrimental substance, perhaps some chemical being
inhaled or eaten, was responsible for the normal aging and death of cells
when part of a multicelled creature. If true, then maybe, just maybe the
chemical could be avoided and death would not be an inevitable part of
life. Immortality, a dream for centuries, might be just around the corner.

Carrel's chicken heart experiment was the scientific news headline of its day. The year-by-year health of the chicken cells was monitored in New York newspapers. False rumors of the death of the

culture in 1940 led the New York *World Telegram* to publish a premature obituary. Science fiction books used the chicken cell culture in plot lines. *The Space Merchants,* written by Frederik Pohl and C. M. Kornbluth in 1953, depicts a monstrous chicken culture named Chicken Little that is tended by numerous workers who carve off vast, renewable slabs to feed a hungry population. If television and *The X-Files* were around during the time of Carrel's experiment, no doubt an episode would have aired where scientists discover the secret of immortality by studying the death-defying chicken cells.

There was just one little problem with these new and exciting ideas on aging and death: the experiment that they were based on was completely erroneous. Yet for fifty years, Carrel's results were viewed as irrefutable evidence that cells isolated from an organism were immortal. If other scientists were having problems keeping cells alive in culture, well then, they were incompetent researchers who must have used dirty glassware or contaminated their nutrients. But no matter how hard the glassware was scrubbed, or how carefully the nutrient broth was made, a growing number of scientists couldn't create cells that were immortal.

Still, a few scientists were successful. In 1951, a thirty-one-year-old African-American woman named Henrietta Lacks had some cells removed from her cervix during a visit to her doctor. The lab that examined her cells reported the worst possible results for the young woman—the cells were malignant cancer cells. Although Henrietta died of cervical cancer eight months later, her cells lived on. Donated to Johns Hopkins University, the cervical cancer cells, code-named "HeLa" for the first two letters of her first and last name, multiplied profusely when cultured in laboratory dishes and were soon distributed to researchers around the world. HeLa cells provided scientists with their first immortal human cell line, a tremendous boon to researchers studying how cells function.[1]

---

[1]Unfortunately, HeLa cells were also the bane of a large number of labs. Many scientists went through the nightmare of discovering that the hamster skin cells (for example) that they began an experiment with mysteriously turned into HeLa cells by the end of the experiment. The trouble with HeLa cells comes from their similarity to tribbles from the *Star Trek* universe; they divide so rapidly that they easily overwhelm other cell cultures if scientists are not extremely careful to keep all HeLa confined to their own containers.

Despite the few success stories, by the late 1950s scientists were becoming increasingly frustrated with their inability to duplicate Carrel's chicken cell experiment—especially when incompetence and sloppiness were the reasons attributed to their failures. Two such scientists were Leonard Hayflick and his colleague Paul Moorhead. No matter how hard they tried, the embryonic skin cells that they placed in culture survived for only a few months. There was also a clear difference between the properties of cells recently removed from animals and immortal HeLa cells. When Hayflick injected HeLa cells into living animals, the cells formed tumors, indicating that HeLa were still cancer cells. This suggested to Hayflick that maybe only cancer cells were immortal. Normal cells taken out of an animal were mortal in cell culture and didn't form tumors when injected into other animals. Although Alex Carrel's conclusions had been accepted for fifty years, Hayflick felt that Carrel had to be wrong.

In any profession, young whippersnappers are discouraged from questioning their elders. When the elders are godlike, Nobel Prize–winning pioneers, the discouragement can take the form of professional suicide. Hayflick knew that he had to be absolutely certain before subjecting himself to the ridicule that would inevitably accompany such dogma-destroying revelations. With each passing year, he conducted more and more experiments that all led to the same conclusion: normal cells were mortal. In one experiment, Hayflick found that only cells that had been dividing for a long time in culture died; cells recently placed in culture were happy and healthy. If dirty glassware or impure nutrients were at fault, why didn't such sloppiness affect all the cells? Hayflick became convinced that cells placed in culture could live and divide for only so long before aging and dying. To test this hypothesis, Hayflick used a single dish to grow female skin cells that had been recently placed in culture and "old" male skin cells (cells that had been dividing for a long time in culture). After a few months, all the cells had two X chromosomes; they were all female. Would a dirty dish kill only the male cells? He reversed the experiment, mixing young male cells with old female cells; now the only cells remaining a few months later were male. Hayflick calculated that cells from embryonic skin when placed in culture could divide about fifty times. After that, they died.

Although Hayflick was convinced that normal cells were mortal, his colleagues at other institutions were not. It seemed unbelievable that Carrel's thirty-four-year experiment could be so wrong. For most of the twentieth century, scientists had accepted as dogma that cells removed from an organism were immortal. So Hayflick gave some young cells to his colleagues and told them to grow the cells in their own labs. He also told them precisely when the cells would die. After growing robustly in culture for four months, the cells given to all the different labs died. To Hayflick, this was the final nail in Chicken Little's coffin. The dream of immortality that arose from Carrel's experiment was just that. Cells were mortal just like the organisms that they came from.

In 1961, Hayflick and Moorhead tried to publish their results. They selected a very prestigious journal, the *Journal of Experimental Medicine*. The paper was soundly rejected. The reason given? That everyone knows cells growing in culture are immortal; if the cells died, incompetent scientists were to blame. Fortunately, the paper was accepted by the journal *Experimental Cell Research* and has since become a classic. Everyone who works on cell cultures today knows of the "Hayflick limit": that normal cells placed in culture divide for a finite number of times and then die. Soon after publication of his paper, Hayflick discovered that cells removed from younger people divide more times than cells removed from older people. Furthermore, cells from animals with long life spans divide more times than cells from animals with brief lives. These results were a revelation. Apparently cells are programmed to divide a certain number of times starting sometime after fertilization of the egg. Hayflick found that human cells cannot escape the fifty-division limit, even if the cells spend part of their lifetime frozen in liquid nitrogen. Cells that have doubled twenty times before being frozen will double only another thirty times once thawed. This led to the idea that there is a clock inside cells, ticking down the number of cell divisions and not simply marking the passage of time. And maybe animals die because a ticking clock inside their cells strikes midnight and signals to the body that it's time to close down.

The idea that a clock ticks down the days of our lives is referred to as the programmed theory of aging. That no matter how well the body is cared for, when one's internal clock reaches its final hour, the party's over. There appears to be an age limit, generally known as the maximum life span, which differs for different organisms and beyond which an organism simply cannot live. For humans, the maximum life span is estimated to be between 115 to 120 years.[2] Of course, the chances of reaching that venerable age are slight today, but much better than they were in years past. During the age of the Roman Empire, the average age you could expect to live to was about twenty-two, which improved to nearly thirty-three years in the Middle Ages, about fifty years in the United States at the turn of the twentieth century, and rests at about seventy-six years for children born today. While modern medicine and a highly technological society are helping people live longer, there has been no increase in our maximum life span. To many scientists, this inability to extend the maximum age implies that our genes determine the time limit of the human body. But this time limit may not be set in stone. There are tantalizing new studies that suggest tinkering with a few genes can vastly extend an organism's maximum life span.

Current research on aging indicates that cells in people rarely, if ever, divide enough times to reach the fifty-division Hayflick limit. Therefore, massive cell death due to the internal cellular clock does not explain the aging of an entire organism. This has led researchers to explore other notions for what controls life span, with exciting initial results. Using fruit flies, researchers recently found a gene that when faulty allows the lucky flies to live a third longer than normal. These "faulty" flies also tolerate high temperatures and starvation conditions much better then their normal counterparts. And just to show that fruit flies aren't all alone in having genes that control life span, a second group of scientists found two genes that when faulty allow worms to live much longer lives, even doubling their normal life span. Worms that contain both faulty genes live up to five times as long as normal, comparable to increasing the human life span to 575 years. At least one of these worm genes is also found in humans.

---

[2]The accuracy of this figure should improve now that most countries keep accurate birth records. Unlike middle-aged people, who tend to shave years off their revealed age, very old people in many parts of the world tend to exaggerate their true age.

Perhaps twenty years from now, the twenty-sixth season of *The X-Files* will have an aging but still youthful-looking Mulder telling Scully that increasing the human life span, like genetic engineering and cloning, was once considered to be science fiction.

Throughout the centuries, man has sought ways to stay young. From ancient times, people have searched for a more perfect food—ambrosia, the food of the Greek gods; fruit from the Garden of Eden; or water from the Fountain of Youth—nourishment that would cure what ails and keep the ravages of age at bay. According to *The X-Files*, the townsfolk of Dudley, Arkansas, have discovered their own Fountain of Youth in the special food that they eat—but unfortunately they dispose of the local food inspector before he can dissuade them from sampling one evening's deadly repast. Dr. Ridley, a disgraced researcher, discovers the secret of immortality while studying children suffering from premature aging—but testing unproven theories on himself and others proves disastrous for everyone. Mulder and Scully probably wish they knew Ridley's secret of reversing the aging process while investigating how a crew and their vessel can age half a century overnight. And for those who plan on burying themselves in frozen capsules waiting for the day when science can bring them back to life, *The X-Files* offers this word of warning—don't mess with the living if you can't guard the plug. While the ability to control aging is still science fiction, much of the underlying science forming the backbone of these episodes is real—or, at least, what is currently held to be true.

Which brings us back to Alexis Carrel. How could his chicken heart cell experiment have been so convincing and yet be so wrong? Chicken heart cells are as mortal as all normal cells, so what kept the cells happily dividing for thirty-four years? During that time, hundreds of scientists tried to duplicate Carrel's experiment with only four success stories. However, all successes were achieved only after scientists treated their cells with chemicals known to change normal cells into cancer cells. Hayflick in his book *How and Why We Age* has a hypothesis for why Carrel's cells never seemed to die. He believes that it was due to a mistake in how the cells were maintained. The nutrients fed to the culture were made from fluid extracted from chick embryos. In that fluid were a number of chick embryo cells. Carrel used a centrifuge to remove the cells from the rest of the fluid—as the centrifuge spun around, cells, which are heavier than other ingredients, should have ended up at the bottom of the test tube. The fluid at the top of the tube was poured off and used to feed the chicken heart cells. Apparently, the primitive centrifuge didn't work very well and some cells remained in the fluid. So every time the chicken heart cells were fed, the culture was provided with fresh cells that could divide until they reached the Hayflick limit! With fresh cells being added to the culture every few days, no one realized that the original heart cells probably died within months of being placed in culture.

While Hayflick was successful at overturning fifty years of incorrect notions on the immortality of normal cells, I would be remiss if I failed to mention that not everyone agrees with Hayflick's conclusions that all cells can divide for a fixed amount of time. One scientist, Harry Rubin from the University of California at Berkeley, thinks that the Hayflick limit is based on artifacts of Hayflick's own experiments—that placing cells into culture is so stressful for the cells that they no longer behave like normal cells attached to a body. In other words, Rubin believes that while the Hayflick limit may be true for stressed-out cells, it may not apply to normal cells. Publishing such dogma-destroying ideas outside of commentary sections of journals may prove very difficult for Rubin and other researchers who share similar views because, after all, everybody knows that cells divide for a finite amount of time and then die.

Turning Back Time

INT. SCULLY's APARTMENT—NIGHT

                    SCULLY
Unbelievable.

                    RIDLEY
Yes. And ultimately irresponsible. My work cost me dearly. I'm
an outcast in the medical community. I was called Dr. Men-
gele, Dr. Frankenstein. But I didn't care—

SCULLY
You knew that if your theories panned out—

RIDLEY
—the man who owns the fountain of youth controls the world.

—"Young At Heart"

In the *X-Files* episode "Young at Heart," federal prisoner John Barnett ceased serving a 340-year sentence when he died of cardiac arrest in 1989. Because of the heinous nature of his crimes, few—and certainly not Fox Mulder—mourned his passing. Years earlier, Mulder's first assignment as an FBI agent was the apprehension of Barnett, but the capture went terribly wrong. Barnett, surrounded by FBI agents, grabbed a hostage, leaving Mulder with a difficult decision— shoot or negotiate. Mulder chose to negotiate. He chose wrong. Barnett savagely killed the hostage and an FBI agent before Mulder finally pulled the trigger. During the trial, Barnett vowed that he would have his revenge on Mulder, the person who tracked him down. Barnett's death left the vow unfulfilled—or did it? Barnett's friend and fellow inmate swore that he saw Dr. Ridley, the physician who signed the death certificate, conducting clandestine surgery on Barnett's diseased right hand on the day of his alleged death.

Four years later, a robbery/murder at a jewelry store shatters Mulder's relief at the passing of Barnett. A note left by the body of the jewelry clerk, written in fresh ink, contains a message identical to one Barnett had previously written for Mulder. Scully, on a hunch, decides to check up on Dr. Ridley. It turns out that Dr. Ridley is not a shining example of the scientific establishment. Ridley is a disgraced researcher, whose medical license and National Institutes of Health grant were revoked when he was found to be experimenting on children without their permission.

By using children in medical research without their consent (or their parents' consent), Dr. Ridley was not only violating the laws of the United States but also the international Nuremberg Code, voluntarily adopted after the atrocities committed by Nazi doctors came to light after World War II. The code implicitly states that only people who are able to comprehend what they are getting into can freely consent to participate as volunteers in medical research. According to the code, Ridley had the responsibility for obtaining the consent of the volunteers himself and not delegating the task to others. Apparently, Ridley did neither.

Abuses of the past—from the Willowbrook State School hepatitis experiments to the Tuskegee Syphilis Study—have led to strict consent requirements in the United States today, and scientists who skirt the rules face the lifetime loss of government grants as well as other punitive measures. It's therefore not surprising that Dr. Ridley lost his medical license and National Institutes of Health grant for experimenting on children without their permission. Unfortunately, the rules for experimenting on people outside the United States tend to be less strict, allowing Ridley to simply pack up his operation and move elsewhere. Although the World Health Organization in 1993 issued guidelines stating that "the ethical standards applied [in the developing country] should be no less exacting than they would be in the case of research carried out in [the sponsoring] country," this is not a rule that is generally followed. For example, in fifteen of sixteen trials in developing countries to examine methods to keep HIV from being transmitted from mother to baby, sugar pills were given to the control group. In similar trials in the United States, all groups had access to anti-viral drugs. While finding less expensive ways to treat women in countries without access to expensive drugs is certainly worthwhile, an editorial published in *The New England Journal of Medicine* suggested that experiments could have been designed to answer the same research questions without the death of many thousands of babies whose mothers received the placebos.

Dr. Ridley no doubt thought that his research was so important that he didn't need to deal with frivolous documents like patient consent forms. Ridley was trying to slow the aging process or even reverse aging, a laudable and lucrative goal. The children that he was studying had a disease called progeria. Progeria, also called Hutchinson-Gilford disease, is one of a handful of extremely rare disorders that seem to mimic premature aging. Children with progeria appear normal at birth, but by the end of their first year it's clear that something in their bodies has gone horribly wrong. The children don't gain much height or weight and begin to develop characteristics usually associated with old age: hair loss, thin wrinkled skin, beaklike noses, joint stiffening, arthritis, diabetes, and heart disease. The children rarely live past their early teens, with most dying of heart failure around the age of thirteen.

The defective gene that causes progeria isn't known. Although scientists are convinced that progeria is a genetic disorder caused by a malfunctioning gene and not some infectious agent, there is never a history of the syndrome in an afflicted child's family. Researchers believe that progeria is an autosomal dominant disorder, but of the rarest type—one that doesn't affect either parent. Recall from Chapter 3 that an autosomal dominant disorder means that the child has inherited a single faulty gene from one parent, and only that defective version of the gene is required to develop the disease. Normally, the parent also suffers from the autosomal dominant disorder, as is the case for Huntington's disease. However, since neither parent of a progeria child has progeria, the disorder is thought to result from a rare, one-in-a-million mutation that spontaneously occurs in an egg or sperm cell. The parent is unaffected, but the child that develops from that egg or sperm has the disease.

In addition to studying children with progeria, Dr. Ridley probably studied a second, more common aging-mimic disease, one called Werner syndrome. Werner syndrome, also known as "progeria of the adult," affects between one in 100,000 and one in 1,000,000 people. Like progeria, Werner syndrome is caused by a single defective gene, but it's a recessive disorder; both versions of the gene need to be faulty to get the disease. Because so few people are carriers (have one copy) of the disease gene, most people with Werner syndrome are the children of parents who are closely related to each other.

Werner syndrome strikes during adolescence. At a time when most teenagers grow rapidly and start being concerned with their looks, these teens stop growing and must be horrified to discover that the reflection in the mirror appears years older than it should, with graying and thinning hair, wrinkled skin and cataracts. People with Werner syndrome are also commonly afflicted with hardening of the arteries and diabetes. Death due to heart disease or cancer usually occurs by the late forties. Dr. Ridley probably switched to studying Werner syndrome because much more is known about the cause of the disease, or at least the gene responsible for the disease.

The hope is always that once a defective gene has been identified, as the Werner syndrome disease gene was in 1996, scientists will be able to understand how a defect in that gene leads to the disorder. This hope has not yet materialized for Werner syndrome. The normal version of the Werner syndrome gene specifies a type of enzyme called a helicase. Helicases are enzymes that "unwind" DNA. Recall that DNA is a double helix—two strands of nucleotide beads wrapped around each other. Helicases unwind and separate the two strands, a necessary event before DNA can be duplicated. There are many different helicase enzymes in cells, and people with Werner syndrome are only missing this one particular helicase. The mystery still remains of how the absence of one of the enzymes that unwinds DNA leads to conditions normally associated with old age.

Recently, though, clues have emerged that just might lead researchers to the answers. Scientists who uncovered these clues were not studying humans with Werner syndrome, but rather tiny yeast cells. Yeast may be more familiar to you as the malodorous powder purchased in little foil packets that causes dough to rise. However, as a single-cell eukaryote, yeast has been very useful for deciphering processes that are important for all eukaryotic cells, including human cells. While it may be difficult to acknowledge that your cells have anything in common with yeast cells, both human and yeast cells need to replicate DNA, make proteins, produce energy, eliminate waste products, and conduct all the other necessary tasks required for cells to live. A recent study found that almost a third of all yeast genes have counterparts in humans.

Yeast cells, like human cells, have a limited life span. So if aging in people occurs wholly, or at least in part, at the level of individual cells, studying how yeast cells age may provide clues for how human cells age. At least this was the theory for why Massachusetts Institute of Technology scientist Leonard Guarente started examining what causes yeast cells to age. He and his colleagues discovered a gene that when defective causes yeast to die much faster than normal.[3] Then the scientists made a really shocking discovery. Guarente searched

[3]Old yeast cells don't turn gray and wrinkled but they are sterile and grow larger than young cells.

for a human counterpart of the yeast rapid-aging gene and found one. The human counterpart is the Werner syndrome gene. Yeast and humans, who haven't shared a common ancestor for at least a billion years, both have a gene that when defective causes premature aging.

Guarente next looked for how the defective gene causes yeast to age and die more quickly. He found something suspicious occurring in the yeast cells' nucleus, the compartment that contains the DNA. Within the nucleus of all eukaryotic cells is a smaller compartment called the nucleolus. Guarente found that the nucleoli in older yeast with the defective gene are swollen and in tatters. The reasons for the misshapen nucleoli seemed to be large numbers of tiny circles of DNA that have detached from the yeast chromosome. Guarente speculates that these little circles, which aren't present in young cells, gum up the workings in the nucleus, causing cells to stop functioning and die. It isn't known whether human cells also make little DNA circles that cause cells to age and die.

Although Dr. Ridley believes that studying children with progeria is a good way to unlock the secrets of aging, not all top researchers in the aging field would agree. According to George Martin, director of the Alzheimer's Disease Research Center at the University of Washington, and Leonard Hayflick, the same Leonard Hayflick who first questioned the immortality of Dr. Alexis Carrel's Chicken Little, children with progeria have only an external appearance of premature aging. While the children may look old and suffer many of the ailments of old age, the precise aging-related diseases that they suffer from are different from those associated with normal aging. For example, only Werner syndrome patients suffer from very rare forms of cancer, like skin cancer on the soles of their feet. Another difference is the distribution of osteoporosis, a disease that causes bones to become fragile and break easily. During normal aging, osteoporosis generally affects the vertebral column, causing a stooped appearance. In contrast, osteoporosis strikes the long bones of the limbs in people with Werner syndrome. Still, finding that the same gene in yeast and humans promotes aging-like symptoms may indicate that Werner syndrome may be a closer mimic of aging than previously thought.

Mulder believes that Dr. Ridley has found a way to reverse the aging process. Although Scully scoffs at his belief as being science fiction, Mulder once again reminds her of the many modern-day advances that were once considered science fiction. Mulder and Scully are informed by another scientist that Ridley believes aging is not inevitable but merely a disease—a disease that could be cured. Scully correctly writes in her journal that there is no evidence that aging is a disease. While aging increases the likelihood of becoming sick from infectious agents, this is because of a weakened immune system. Infections by themselves do not cause aging.

Ridley has apparently learned that Scully is making inquiries about him and appears on her doorstep to provide his side of the story. Ridley tells Scully that the goals of his research are, as Mulder suspected, to halt or even reverse the aging process. Ridley mentions that his methods involve the use of gene therapy. Since the gene responsible for Werner syndrome has been isolated, Ridley may have tried using this gene to stop the aging process. Perhaps his hypothesis was as follows: because mutations in the Werner syndrome gene lead to Werner syndrome and premature aging, maybe extra copies of the normal gene would prevent aging. In other words, Ridley knew that people with Werner syndrome fail to make the helicase enzyme in their cells. If lack of the helicase leads to aging (maybe because the helicase keeps the little DNA circles from forming in cells), then maybe extra amounts of the helicase might keep people from aging. Ridley may have thought that by providing his cells with extra copies of the Werner syndrome gene he would increase the amount of helicase enzyme in his body and consequently would cease, or even reverse, the aging process.

Dr. Ridley may therefore have attempted to use gene therapy to insert extra copies of the Werner syndrome gene into his own genome and the genomes of his experimental subjects. This scenario assumes that Dr. Ridley was able to perfect a virus vehicle to taxi the new genes into all or most of his cells. This little problem aside, the consequences of the gene therapy to his body are clearly disastrous. Instead of reversing the aging process, Ridley has developed terminal vascular disease, a common symptom associated with progeria and Werner syndrome. Only one of Ridley's patients, the evil Barnett, regained his youth; the others only gained an early grave.

What could have gone wrong with the gene therapy? Somehow, with the exception of Barnett, Dr. Ridley has ended up achieving exactly the opposite of what he was hoping for. Instead of halting aging, he has developed symptoms of early aging. Conducting an experiment and getting the opposite of what you hope for is not an unknown event in science. If Ridley was trying something as experimental as gene therapy with the Werner syndrome gene, he may have discovered that inserting extra copies of a gene into his genome doesn't necessarily lead to a cell making extra protein specified by the gene.

It may seem obvious that when you add more of something, you get more of what you want—in the same way that adding two cubes of sugar will make your tea sweeter than one cube. If, however, you plug twenty lamps into a circuit to get twenty times as much light as the single lamp now on that circuit, oops, the circuit overloads, causing a fuse to blow, leaving you with no light and possibly no house. And biology is much more complicated than electrical engineering. This may be what happened to poor Dr. Ridley. Instead of adding extra copies of the Werner syndrome gene and making more helicase enzyme in his cells (which may or may not have any kind of beneficial effect on aging), he may have inadvertently caused a situation where he blew the circuit on his own normal Werner syndrome genes, leaving him with no helicase enzyme and, consequently, no life.

This is not pure conjecture on my part. A while back, some scientists had the lofty goal of trying to make purple petunia plants with more vibrantly colored flowers. Their hypothesis was that if they added extra genes that specified a protein involved in purple flower color, they could create plants with richer-colored flowers. The experiment should have worked. However, instead of getting plants with dark purple flowers, the plants genetically engineered to contain extra purple color genes had white flowers—flowers with no color pigment at all. Somehow, putting extra copies of a gene into the plant had caused the new genes and the plant's natural color genes to shut down completely, hence the flowers lost all pigment. Why the plants turned off their own flower color genes is still a scientific X-File. But whatever caused the mysterious gene turn-off in the petunia plants (and in many other plants subjected to similar experiments) may also have caused Ridley's experiment to backfire. Ridley may have inadvertently shut down his own Werner syndrome genes by adding extra copies of the Werner syndrome gene to his cells. And without his own helicase genes functioning normally, Ridley would be in the same boat as people who suffer from Werner syndrome: he would stop making the helicase protein. This would cause Ridley to come down with the disease and vascular problems associated with Werner syndrome . . . and die.

Diseases such as progeria and Werner syndrome have helped to eliminate one of the theories of aging—that aging and death occur because the body simply wears out after many years of hard use. This theory, popularly known as the "wear and tear" theory, was first suggested over one hundred years ago by German biologist August Weismann. Weismann believed that tissues eventually lose the ability to renew themselves. The theory suggests that the wear and tear, the stress, of everyday life exacts a toll on the body's vital systems that simply wears them out. However, since children and young adults with aging-mimic diseases have hardly lived long enough to suffer wear and tear, this theory cannot account for their aginglike symptoms and therefore has few proponents today.

Any theory on why organisms age has to take into account the Hayflick limit—that individual cells divide a limited number of times and then die. Scientists were interested in determining whether the mysterious clock that ticks off the number of cell divisions differed for people suffering from progeria and Werner syndrome. Were cells from these patients permitted fewer cell divisions than normal cells before dying? The answer was yes. Not only did progeria and Werner syndrome patients have a condition that seemed to accelerate the aging of their bodies, but their individual cells aged more quickly as well.

Ever since Hayflick suggested that some mechanism within cells controls aging by ticking off each cell division, scientists have searched for the clock. In 1972, American Nobel laureate James Watson and Russian Alexei Olovnikov simultaneously came up with an intriguing idea. Each time a cell divides, it needs to duplicate its DNA. What if each time DNA is duplicated, the newly made DNA is a bit different from the original DNA? Recall that the DNA in a chromosome is one long double helix, like a long piece of braided string strung with nucleotide beads. A perfect duplicate of this DNA string would be exactly the same length as the original DNA. Yet with their fledgling understanding of how enzymes in cells duplicate DNA, both scientists independently realized that newly made DNA couldn't be exactly the same length as the original DNA.

Here's why: Imagine there is a line drawn from wall to wall in one room of your home. Your task is to duplicate that line. However, you must hold a crayon in your toes and you can only walk continuously in one direction. Your best bet, therefore, is to stand against one wall and walk straight across to the other wall, drawing the line with the toe-held crayon. The duplication of the original line is perfect (providing you don't trip) except at the very beginning of the new line, where a distance the length of your foot isn't copied. The new line therefore has shrunk by the length of your foot. Now copy the line you just made by going in the opposite direction; this requires that you begin by standing against the opposite wall. Again, the new line will be shorter than the one you are copying by the size of your foot. Each time the DNA in a chromosome is copied, about one hundred to two hundred nucleotide beads are lost from the ends because the enzyme cannot start at the very beginning of the DNA. Alexei Olovnikov, who was familiar with Hayflick's clock idea, thought that this DNA shrinkage could be the clock—that every time a cell divides, the ends of the DNA shrink by a small amount. And once the DNA has shrunk to a certain size, some signal is sent out that tells the cell to die.

Since this original idea was bandied about, evidence has been pouring in implicating the shrinkage of the ends of DNA as the cell's timepiece. First, DNA ends were found to be very different when compared with the rest of the DNA. The order of the nucleotide beads at the ends of DNA in chromosomes is of the "junk" and not gene variety, but very unusual junk—tens of thousands of copies of a particular order of six nucleotide beads. The ends of chromosomes were even given a name: telomeres. Then came some evidence that connected the length of telomeres with the age of a cell. Cells that had been dividing for a long time in culture had shorter telomeres than cells just placed in culture, and telomeres from skin samples taken from older people were shorter than those from the skin of young people.

Imagine every time you tie your shoes, the caps at the ends of your shoelaces become a little shorter. A point would come where the caps disappear and you would probably toss the laces. If the cosmic shoelace-cap fairy waves a magic wand and regrows the caps each time you tie your shoes, the laces will last for a much longer time. If the laces were protected from harmful substances, they might even last forever.

Telomeres at the ends of DNA are like shoelace caps. Each time a cell divides, the telomeres shrink a bit in size. While they never shrink away completely, a point is reached when a message is sent to the rest of the cell that tossing time has come. But what about cells that are immortal, like cancer cells? When scientists examined cancer cells, they found that cancer cell telomeres don't shrink each time the cells divide. Apparently cancer cells have a cellular telomere fairy who regrows the telomeres after each cell division, keeping cancer cells alive forever.

The identity of the cellular telomere fairy was revealed in 1984 by Carol Greider and Elizabeth Blackburn. These scientists were not studying how cancer cells stayed alive but rather why telomeres don't shrink in a tiny, single-cell eukaryote called *Tetrahymena*. The scientists discovered that *Tetrahymena* make an enzyme that can lengthen the telomere caps on the ends of the DNA by replacing nucleotide beads that are lost when the DNA is duplicated. They named the newly dis-covered enzyme telomerase. When scientists searched human cells for the telomerase enzyme, they didn't find it in normal cells. In contrast, the telomerase enzyme was found in nearly all types of cancer cells. This discovery led to a new method for the early detection of cancer: the identification of cells that have the telomerase enzyme.

If the telomerase enzyme keeps cancer cells alive by not letting telomeres shrink, the next question facing scientists was what would happen if they provided normal cells with their own telomerase fairy? Would the cells become immortal? Would they turn into can-cer cells? The answer to these questions came in early 1998. Scien-tists at Geron Corporation, a California biotechnology firm, used genetic-engineering techniques to give normal eye cells and skin cells their own personal telomerase fairy. Following the genetic en-gineering procedure, the cells were returned to culture, where they began to divide ... and divide ... and divide. When the normal fifty-division Hayflick limit was reached, the cells not only didn't die but they maintained the appearance and features of younger cells. Aging for these cells had apparently stopped. At the time that I write this, the cells are still youthful and dividing.

The discovery that cells have a timepiece that ticks off a cell's life span has important implications for cloning people. If the cell that provides the nucleus for cloning comes from an older person, say an individual who is sixty years old, the ends of the DNA inside that cell will already be considerably shrunk. Will the baby cloned from that old nucleus start life with the DNA of a sixty year old? How will this affect the life span of the child? Unfortunately, experiments in other mammals may not give answers applicable to humans since it is known that the telomeres of mice have properties quite different from the telomeres of people. This means that until a child is cloned and can be monitored, it won't be known if the cloning has vastly reduced the child's maximum life span (and the life span of any children of the cloned individual). Subjecting these children to the status of living guinea pigs without their consent is ethically indefensible.

When Geron released the news that they had increased the life span of cells in culture by providing them with telomerase, headlines proclaimed their discovery as an attack on aging and as a possible fountain of cellular youth. People with visions of immortality drugs caused Geron stock to shoot up 44 percent overnight. Yet as Geron scientists quickly pointed out, all they had done was keep some cells alive in culture for longer than normal. The only entities imbibing from their Fountain of Youth were individual cells confined forever to plastic bottle cages.[4] And while this study received substantial pub-

---

[4]While the Geron experiment of providing telomerase enzyme to normal cells is unlikely to be the Fountain of Youth as hyped in the press, an important finding did emerge: normal cells that produce telomerase become immortal without developing any of the abnormalities of cancer cells. The telomerase fairy therefore doesn't turn normal cells into cancer cells, it just keeps cancer cells alive. The implications of these findings are very exciting for cancer therapy. Recall that scientists are constantly searching for differences between cancer cells and normal cells. Since almost all cancer cells make telomerase while normal cells do not, a new idea in cancer therapy has emerged. What would happen to cancer cells if their telomerase fairies are destroyed? Would these rapidly dividing cancer cells quickly reach the Hayflick limit and die? The Japanese, who have very low rates of cancer, may have already provided the answer. Scientists in Japan have discovered that a chemical in the green tea enjoyed by so many Japanese naturally blocks the ability of telomerase enzyme to function. They suggest that this might be a major mechanism underlying the documented anti-cancer effects of green tea. Human trials to test the efficacy of destroying the telomerase enzyme in cancer cells will soon begin.

licity, a later study revealed that not all types of cells become immortal simply by adding telomerase. For some cell types, a second gene that normally functions in cells to suppress the formation of tumors must also be inactivated. Clearly, achieving immortality even for a single cell is more complicated than previously thought.

The best accepted theory for what causes organisms to age, the free radical theory, is presented in the *X-Files* episode "Dod Kalm." In this episode, Mulder and Scully journey to the coast of Norway to investigate how American sailors picked up in a lifeboat appeared fifty years older than their true age. When Mulder and Scully reach the sailors' former ship, the USS *Arden*, they find a rusting hulk that is no longer seaworthy and a thirty-five-year-old captain near death from apparent old age. Abandoned on the ship, Mulder and Scully find themselves aging as well, with little time to find the cause or cure.

Mulder's theory for why he, Scully, and the ship are aging involves a complicated concoction of wormholes developed using alien technology and a government conspiracy that has selected the North Atlantic for experiments in time displacement. Scully, naturally, prefers a somewhat less theoretical theory. She believes their aging is due to the presence of toxic levels of chemicals known as free radicals. Scully explains to Mulder that free radicals are highly reactive chemicals that can attack DNA and proteins and are the leading suspects in why our bodies age.

Scully is describing what happens when an insidious chemical gets loose in our bodies. This poisonous element causes widespread mayhem by damaging proteins and DNA inside cells. If the free radical theory is correct, then aging should be "curable" by simply avoiding the ingestion of the treacherous chemical that causes the formation of free radicals. So what is the chemical to be avoided if we are to achieve immortality? Oxygen. A sad reality of life is that avoiding oxygen is not recommended for longevity. A decision made billions of years ago by ancestral prokaryotes to use oxygen for energy production is now the bane of highly evolved humans. The same oxygen that keeps our bodies alive may also be the cause of our ultimate downfall.

Oxygen in the atmosphere is harmless. But like a school yard bully, oxygen gets nasty in the presence of "weaker" molecules, molecules that have trouble holding on to their electrons. The problem with oxygen is that it likes to grab electrons from other molecules, much like a bully grabs lunch money from a weaker child. Metals are particularly vulnerable to greedy oxygen. When oxygen grabs electrons from a metal like iron, the iron is said to be oxidized and it rusts. While no one likes to see the metal in their car or bicycle rusting away, a more pressing concern should be what happens when oxygen plays its little grabbing games inside one's own body.

Although few of us will ever be caught gnawing on a bicycle, metals still make their way into our bodies and cells. When oxygen grabs an electron away from a metal, the oxygen becomes a highly destructive molecule known as a free radical. Just like a bully who finds it easier to take money from other victims after achieving success with the first, free radicals with their stolen electron find it much easier to steal an additional electron. This is because atoms are more content when their electrons are present in pairs of two. Oxygen with an extra unpaired electron can violently grab an additional electron, even from a molecule that doesn't want to give one up. When oxygen grabs an electron away from another molecule, the victim molecule now has one less electron than it needs, so it will grab an electron away from some other poor molecule, and so on. This chain reaction can be highly destructive, especially if an electron is grabbed away from proteins or DNA. Free radical–damaged proteins cause skin aging and cataracts. If DNA is the target of the free radical's electron grab (as it frequently is), damage can lead to breaks in the DNA strands and create mutations that can also start the cell down the path toward cancer.

Many scientists believe that damage to the cell's DNA caused by free radicals is one of the root causes of aging. This view gained even more supporters recently when a team of Japanese and American scientists showed that the life span of worms is reduced when a particular gene becomes mutated. The gene specifies a protein that helps to keep free radicals from forming. When the gene is mutated, more free radicals are made and the worm lives a shorter life. Place the mutant worm in a container with a high concentration of oxygen, and it dies even sooner.

Free radical bullies do more than just promote aging and cancer. Evidence is accumulating that free radical reactions are the underlying cause of heart attacks, strokes, and Alzheimer's disease. While free radicals produced from breathing oxygen cannot be avoided, other sources of free radicals are less natural: air pollution, tobacco smoke, radiation, and toxic waste. According to many scientists, taking steps to reduce air pollution, avoid cigarette smoke, preserve the ozone layer to shield the sun's radiation, and clean up toxic waste dumps would have an enormous benefit for everyone's health.

Fortunately, organisms that have to breathe oxygen are not completely at the mercy of oxygen's destructive nature. Cells contain enzymes that hunt down and destroy free radicals. But glance at your wrinkles in the mirror as you get older and you will notice that these enzymes don't do a first-class job. What's needed are chemicals that will unselfishly give up an electron to a free radical without becoming a link in an electron-grabbing chain. Chemicals such as these should be able to protect cells from free radicals, and help prevent heart disease, strokes, and cancer.

Such chemicals are not fiction; they exist and you've probably heard of them. They are called antioxidants. Vitamin E found in wheat germ, oils, nuts, and fish; vitamin C found in citrus fruits, broccoli, green pepper, and spinach; and beta carotene found in squash, cantaloupe, carrots, and yams are all antioxidants. There are many scientific experiments that point to antioxidants as being important for longevity. Mice that are fed large doses of antioxidants live 30 percent longer than mice fed a normal diet; fruit flies that live up to 60 percent longer than the average fly have higher amounts of antioxidants in their cells. There is even a connection between organisms that live the longest and high natural levels of antioxidants. Is it any wonder that the USDA food pyramid recommends all those servings of bread, fruits, and vegetables?

Free radicals are not always the villains in the saga of life. Your white blood cells, called phagocytes, prowl around the bloodstream looking for foreign bacteria and other microbes to eat. Phagocytes dispose of these invaders of your body by placing them into little compartments inside the phagocytes that contain high levels of free radicals. Unfortunately, prolonged infections cause the phagocytes to expire from too many meals of bacteria, and their death releases all those free radicals into the bloodstream. The damage caused by these free radicals is serious, and can lead to painful conditions such as rheumatoid arthritis.

Scully believes that the USS *Arden* is being bombarded with extremely high levels of free radicals, causing the rusting of the hull and the aging of people on the ship. She has a fascinating theory for what might be producing all those free radicals. Scully believes that the ship is drifting toward another massive metallic object, such as a buried meteor. The ship and the meteor could be acting as positive and negative terminals with the salty ocean acting as a giant battery. Unfortunately, Scully's theory breaks down because of the large expanse of ocean lying between the two objects. To produce an electromagnetic field strong enough to generate oxygen free radicals, the objects (ship and meteor) would have to be only a fraction of an inch apart. If massive doses of free radicals are the cause of their aging, then something else must be the catalyst.

While Scully doesn't realize the flaws in her theory, she soon finds that free radicals are not to blame for the aginglike symptoms. The water that comes from the sea through the desalination system is somehow producing extremely high levels of sodium chloride (table salt) in the body, and it is the salt that is causing the substantial cellular damage that resembles aging. Fortunately, Scully is aging slower than Mulder and is able to continue her scientific analyses of their condition. Scully's ability to better withstand the affects of the salt are attributed to the general difference in aging between men and women, with women living longer than men.

Even before women are born, they have a lower death rate than men. Up to 170 boys are conceived for every one hundred girls but a disproportionate rate of spontaneous abortions, stillbirths, and miscarriages of male fetuses results in an almost identical number of boy and girl babies born. At every stage in a girl's life, her life expectancy outstrips that of boys. Some scientists have attributed the difference in life expectancy to the difference in sex hormones. The onset of puberty and the increased production of testosterone in boys leads to mortality from reckless behavior or violence that is not confined to humans; primates like macaques also have a similar surge in death at a comparable stage in life.

While men eventually finish sowing their wild oats, the difference between male and female mortality increases once more at early middle age, when heart diseases tend to strike men at least ten years before women. Some scientists speculate that testosterone is again the culprit since it increases levels of the bad cholesterol (known as LDL for low-density lipoprotein) and decreases the levels of good cholesterol (HDL, high-density lipoprotein). Estrogen, the female hormone, has the opposite effect on cholesterol, and may therefore protect women from heart disease. Estrogen also has another property that lessens the diseases that women get—it's an antioxidant. Estrogen therefore helps to placate those unpleasant free radicals that cause so many lethal diseases. While women have fewer fatal diseases than men, they suffer from more nonfatal conditions such as arthritis and osteoporosis. Women also suffer the side effects of having a slower metabolism than men, with many having an endless battle with weight control. However, studies with animals have shown that restricting food, which slows the metabolism, increases longevity by as much as 30 percent. A slow metabolism, while a curse during much of many women's lives, may therefore be a boon in the twilight years.

Fortunately for Scully, a woman's slower metabolism kept her alive long enough to complete her observations on the problems she and Mulder were facing. After a timely rescue by the navy just minutes before the rusty ship sinks, they are quickly treated at a hospital based on Scully's notes. Happily for fans everywhere, Mulder and Scully regain their youthful appearance after the treatments, indicating that they were suffering from a condition that only mimicked aging. Sadly for the people with progeria and Werner syndrome, their aging-mimic diseases are not reversible and nothing yet tried slows down the progress of these diseases.

Aging is an enormously complicated issue. There are as many theories as to why we age as there are scientists working in the field. Whether aging is programmed in our genes or is a lifetime's worth of accumulated cellular damage due to free radicals, or perhaps some mixture of both, it is an inescapable part of living. Whether the cellular clock is telomere shrinking, generation of tiny DNA circles, or some still unknown attribute, only time will tell. And whether this work will expand the human life span is only conjecture at this point. Yet research on aging in yeast, fruit flies, worms, and humans may very well lead to novel treatments for cancer, arthritis, heart disease, and other maladies of aging, which should translate into a better and longer life for us all.

Freezing Time

It's a tempting thought. Have your body placed in suspended animation and wait for the scientific advancements that will cure your ills. Over seventy patients currently lie frozen in capsules waiting for a remedy to combat the illness that they all share. That illness is called death—at least death as it is currently defined. Just as drowning victims who would have been pronounced dead one hundred years ago today could have been resuscitated if their time under water wasn't overly long, those who place their hopes in the freezing sleep of the dead called cryonics expect that today's definition of death will not encompass all future definitions of death. Arthur Grable, an engineer at the Mahan propulsion laboratory in the *X-Files* episode "Roland" must have believed in the Cryonics Society motto: "Freeze—Wait—Reanimate." He, like hundreds of others, signed up to have his body frozen after death. It's not cheap. For a fee ranging up to $150,000 for a full body and $50,000 for head only, plus yearly dues to cover subscriptions to quarterly magazines and members' newsletters, Arthur arranged to have his body "donated to science" upon his death—the normal way that cryonics companies acquire patients. Arthur knew that if he encountered a fatal mishap or when his body finally succumbed to disease or organ failure, the cryonics company would see to it that his coffin would be a stainless-steel capsule with icy-cold liquid nitrogen provided in lieu of flowers.

Arthur was not planning on dying young. He probably thought that he had plenty of years to develop the next generation of jet engines. Doubling engine speed using only half the current fuel would endow him with fame and fortune beyond the dreams of most engineers. Arthur was very close to discovering this holy grail of engineering when a sudden car accident ended his life. Arthur, however, was fully prepared for this possibility. Like all those who have signed up for cryonic suspension, Arthur must have carried identification tags identifying him as a cryonics candidate. The tags provide instructions with handy tips for properly storing the body until it can be reached by trained emergency stabilization teams from the cryonics organization.

After Arthur's body temperature was reduced to just above freezing, his blood was replaced with an antifreeze agent, followed by a five-day descent into the frigid conditions of -320 degrees F. Arthur's remains were then stored within an extremely well-insulated container similar to a household thermos bottle (only a bit larger). Occupying space inside the container with Arthur is the liquid nitrogen needed to keep his body frozen for decades, centuries, or millennia. During that time, Arthur would be perfectly preserved. The amount of tissue degeneration that occurs during one second at normal temperatures would take 30 trillion years at -320 degrees F. Radiation is a bit larger problem than tissue degeneration, but still would take about sixty thousand years before an already dead Arthur would accumulate a normally lethal dose.

One consequence of Arthur's untimely accident is that his promising experiments are taken over by the remaining scientists on the research team. Most scientists would want others to continue their work after death or a less fatal departure from a laboratory; otherwise, the project dies and all effort spent is wasted (I am speaking from some experience here). Unfortunately, those continuing with Arthur's work decide to also take credit for his ideas, a not uncommon occurrence when a scientist leaves a laboratory. Still, with Arthur dead, his colleagues probably thought that he wouldn't care.

But somebody did care. Six months after Arthur's death, team members taking credit for Arthur's research start dying. Furthermore, someone is accessing Arthur's computer files and new theoretical equations are mysteriously appearing on the scientists' blackboard. When Mulder and Scully are called in to investigate, they determine that only one person was present at the time of the first researcher's death—Roland, the mentally challenged janitor hired by the late Arthur Grable. While Roland has an uncanny ability to count objects, he seems an unlikely candidate for the source of the complex fluid dynamic equations. While Roland had the opportunity to commit murder, he had no apparent motive. Mulder and Scully therefore propose a second possibility: Arthur Grable faked his own death. When the FBI agents present this second theory to the remaining scientists, they are taken to the cryonics facility to view the stainless-steel capsule containing the very dead Dr. Grable.

Cryonics, the latest craze in life extension, was born in 1964 when Robert Ettinger, a Michigan physics teacher and author of the book *The Prospect of Immortality,* wrote, "We need only arrange to have our bodies, after we die, stored in suitable freezers against the time when science may be able to help us. No matter what kills us, whether old age or disease, and even if freezing techniques are still crude when we die, sooner or later our friends of the future should be equal to the task of reviving and curing us." Within a few years, the word *cryonics* was coined and several organizations started offering the service. On January 12, 1967, Dr. James H. Bedford, a seventy-three-year old retired psychology professor from Glendale, California, became the first to enter the deep freeze after his death from cancer. Contrary to popular beliefs, he was not joined in frozen sleep by Walt Disney. As I endlessly explained to people taking my tours when I guided visitors at Disneyland, Walt is not lying frozen under the Pirates of the Caribbean. He was cremated and buried in 1966.

When Mulder and Scully view what is left of Dr. Grable they are shocked to find that only his head is frozen. When they point out to the person minding the freezers that living as a disembodied head might affect Arthur's future quality of life, they are reminded that if a time comes when scientists are actually able to thaw a brain and return it to life, cloning a body for the head should be child's play.

The cryonics caretaker is not mistaken when telling Mulder and Scully that science is going to have to advance considerably before reviving frozen corpses—with or without a body. The science of cryonics is still in its infancy. Anyone can pop a living or nonliving head into liquid nitrogen, as is visually demonstrated on one of Arthur Grable's colleagues in the episode. However, thawing out a head whose cells even remotely resemble the original has about as much chance of succeeding today as reconstructing the colleague's head after it shatters into hundreds of pieces when contacting the hard laboratory floor.

There is no question that individual cells can survive being frozen and thawed. David Gilichinsky, a microbiologist at the Russian Academy of Sciences in Moscow, extracted 3-million-year-old bacteria embedded in ice below the frozen surface of Siberia. Within hours of being thawed, the bacteria started dividing. In 1949, Audrey Smith, Christopher Polge, and Alan Parkes at the National Institute for Medical Research in London were the first to show that animal cells could be kept alive following freezing and thawing provided that chemicals called cryopreservatives were added to protect the cells. Today many single cells and simple tissues such as eggs, sperm, white blood cells, skin, cornea, and pancreatic islets have been frozen for long periods of time. Frozen embryos thawed eight years after freezing have gone on to become healthy humans.

Problems arise when trying to freeze cells that are still connected to large, multicellular warm-blooded mammals, like humans. There are two major difficulties that need to be overcome before any mammal can be successfully reanimated after being frozen. The first problem is the massive tissue damage that occurs during the freezing process. As body temperature sinks, water is sucked out of cells, leading to dangerous dehydration. At temperatures just above freezing, water molecules both inside and outside of cells start forming jagged icicles that pierce cells, turning tissue into mush. As if the problem with ice crystals isn't enough, the membranes that surround every cell, being composed mainly of fat molecules, do what all fats do when cooled—they stop being fluid liquids and become solid (think about bacon drippings cooling in a pan). This causes cells to shrink, which creates cracks and crevasses in the membranes that let the cellular contents ooze out. Since organs contain many different cell types, each with a slightly different composition of fats in their membranes, the cells in organs freeze and shrink at different rates, causing a chaotic tug-of-war between cells in the organ. Thawing creates even more damage, as ice crystals get even larger at temperatures just below the freezing point.

And that's just problem number one. Problem number two is even worse. This is the predicament called ischemia, which is defined as the damage to tissues resulting from oxygen and nutrient deprivation. When a person dies, the heart stops, which shuts down the circulatory system. Many cryonicists believe that the body is still 99 percent alive at this time with most of the tissues, cells, and functions still intact—but not for long. Without a functioning circulatory system, cells in the body no longer get the oxygen that they need to survive. Without oxygen, cells can't make their ATP fuel, and without fuel, cells function about as well as cars do when running on empty. Brain cells are especially hard hit because fuel is needed to keep their electrolytes in balance. If ions like sodium, calcium, and potassium are not kept in their proper places inside and outside of brain cells then all the biological machinery in the cells starts breaking down. Much of the harm from freezing damage comes from cracking. Like a crack in a window, the pieces are still there ready to be fixed by future technology. Ischemia, however, causes the destruction of proteins, DNA, and other cellular molecules. Reconstructing such damage, even with future technology, will prove difficult if not impossible.

Cryonics companies are well aware of the damage that occurs with ischemia. Minimizing this damage depends on quickly freezing the patient before the cells react negatively to the loss of oxygen. Cryonics companies have teams in many major cities in the United States that rush to the sides of dying or dead clients. Still, time is short—maybe thirty to sixty minutes—before widespread cellular destruction occurs. Most people who sign up for cryonics suspension, like Arthur Grable, don't have the "ideal" death in a hospital with teams from their cryonics company standing by. Consequently many people with dreams of experiencing the future never end up being frozen in the first place and must settle for a more traditional burial.

The key ingredient in cryonics treatment is the cryopreservant. Cryopreservants are the antifreeze agents that are quickly pumped into a cryonics patient to replace the blood. The composition of the cryopreservant is crucial for minimizing the formation of ice crystals and membrane cracking that occurs from the freezing procedure. Cryopreservant formulas are based on studies of animals that naturally cope with cold conditions. Many cold-blooded animals produce large amounts of sugars when the temperature drops, which help to reduce cellular dehydration. Spiders, ticks, mites, and polar marine fish can survive with body temperatures as low as 5 degrees F by making special antifreeze proteins that are covered in sugars. These proteins have two functions: they coat ice crystals as they form, preventing tiny crystals from growing larger and damaging cells, and also lower the temperature at which freezing occurs.

Some turtles and frogs can endure weeks of having over half of their body water frozen. Wood frogs accomplish this amazing feat in two ways. First, when temperatures drop, the frog's liver puts out vast quantities of the sugar glucose, which gets picked up by other cells. Having so much glucose inside cells lowers the freezing point of the cell's liquid, thus preventing ice crystals from forming. Still, if it is cold enough, icicles will form so the frog has a strategy to minimize how large they can get. Frogs make proteins that actually help ice crystals to form. While this might seem to be counterproductive, by producing large numbers of these proteins in the blood, so many little ice crystals form that can't get large enough to cause damage. The second benefit that nature has provided to frogs is evident when the frogs thaw in the spring. The frogs thaw from the inside out, in the opposite direction of freezing. This allows the organs that were last to freeze to be the first to thaw; the heart starts beating before the limbs thaw, allowing the frogs to thaw without ice crystal damage to their cells.

Unlike cold-blooded frogs, mammals are very sensitive to freezing. If even 2 to 3 percent of a mammal freezes, it dies. With no natural ability to manufacture cryopreservants, mammals have developed their own means of surviving cold temperatures—warm fur and down parkas. Scientists have also been exploring methods to keep mammals from dying when their body temperature drops. One method is to replace the blood with cryopreservant solutions containing the same types of antifreeze agents that keep cold-blooded animals alive under similar conditions. The key is finding a concoction of animal antifreeze substances that work best, a subject of considerable research today. Nevertheless, while dogs and baboons can survive many hours with a body temperature below 68 degrees F if a suitable cryopreservant liquid replaces their blood, no mammal or even frog can survive being completely frozen, even for a short amount of time.

According to the *X-Files* episode "Synchrony," there is one more use of cryobiology—protection from the extreme heat of time travel.

```
INT.–MULDER'S CAR–NIGHT–MOVING
```

```
                    MULDER
Physicists like Stephen Hawking have hypothesized the ex-
istence of worm holes and closed-time-loops—actual portals
through which matter can travel backward through time. But
phenomena like extreme heat and gravity would presumably
render the trip lethal for any organism.
```

                    N I C H O L S
And you're saying the properties of my compound will make it
possible—

                    M U L D E R
                      (shrugs)
    Eventually.

                    N I C H O L S
But why stop time travel?

                    M U L D E R
That's what I'm hoping he can tell us.

                                        —"Synchrony"

    A combination cryopreservant and rapid-freezing agent devel-
oped by young Dr. Nichols of the Massachusetts Institute of Technology leads to such anarchy in the next century that old Dr. Nichols
returns through time to stop his younger self and colleagues from
discovering the chemical in the first place. Old Dr. Nichols first con-
fronts a Japanese colleague, Dr. Yonechi, who is an expert on vitrification. Vitrification, or "glass making," is a real scientific method of
supercooling cow and human embryos, ova, red and white blood
cells, and corneas without forming ice crystals that tear cells apart.
The idea behind vitrification is to combine high levels of cryopreservants with rapid cooling so that ice crystals don't have time to form.
Cells simply lose heat until a point is reached where the liquid inside
cells "locks up" into a solid state. Substances that are vitrified look
like clear glass but feel like hard putty. When the vitrified cells or tissue is rapidly thawed and the cryopreservant washed away, there is
no residual damage from the prior freezing.
    Old Nichols explains to a startled Dr. Yonechi that Yonechi will
solve the problem of vitrification of entire organisms, and for this he
must die. When trying to vitrify entire organs, the large number of
tissue layers makes it impossible to get the cryopreservant to permeate through to the inner tissue. This is analogous to dropping a chunk
of hard bread into a pail of cryopreservant water; even after waiting

a reasonable period of time, only the outer layers of the bread will be wet. Cut the bread open, and no water is in the center. Even if the cryopreservant reaches all tissue layers, rapid freezing is also essential to achieve vitrification. All the cells of the organ must uniformly reach the same temperature at the same time, which isn't possible today. Rapid and uniform thawing of vitrified tissue is also required for the organ to remain healthy; the trouble is, as anyone who eats ice cream knows, the outsides of a frozen substance thaw quicker than the insides.

Old Dr. Nichols tells Dr. Yonechi that he will solve the problem of vitrification by developing a cryopreservant agent that not only permeates organs quickly and uniformly, but causes instantaneous freezing of the entire body. Scully runs an analysis of the substance using a technique called nuclear magnetic resonance, or NMR. NMR allows scientists to identify molecules and determine their structures. Lisa, Dr. Nichols's research assistant, looks at the NMR profile of the substance and recognizes it as a theoretical molecule that she and Dr. Nichols have been working on. Lisa explains to Scully that the substance acts like a catalyst and initiates a "self-sustaining endothermic reaction." In other words, the substance quickly permeates the body and then causes all the heat in the body to be absorbed. Without heat, only cold remains and the body quickly vitrifies.

Such a substance is, of course, pure science fiction. Freezing an entire organism with current technology is slow and uneven, resulting in major tissue damage when the organism is finally thawed. Dr. Grable, as an engineer, would have been well aware of the problems associated with cryonic preservation. So why do intelligent people like Dr. Grable pay cryonics companies a small fortune for what is likely to be a very expensive coffin? Believers in cryonics would answer that question with one word: nanotechnology.

Everything around you and inside you is made of atoms— atoms like oxygen, nitrogen, phosphorus, carbon, and about ninety-nine other elements. If it were possible to build objects atom by atom, then anything from proteins to DNA to the space shuttle could theoretically be constructed from atoms that are naturally found in water and dirt. Imagine that single Lego building blocks are individual atoms. Just as a child can assemble a variety of objects using the same set of Lego blocks, so too might just about anything be assembled from basic atom building blocks. The only stipulation is that the assembly cannot conflict with the natural laws of the universe.

Nanotechnology is the manufacturing of virtually anything, starting with individual atoms. The theoretical possibility of nanotechnology was first voiced in 1959 by Nobel laureate Richard Feynman, who stated, "The principles of physics, as far as I can see, do not speak against the possibility of maneuvering things atom by atom." Of course, assembling items from individual atoms is going to take incredibly small machines that are extraordinarily precise. Eric Drexler, who popularized the subject of nanotechnology in his Ph.D. thesis, has proposed the possibility of tiny machines called assemblers. The size of an assembler would be about 100 nanometers (hence the name "nano" technology), which is about four ten-millionths of an inch. The assembler would have a tiny molecular computer and a robot arm that would be able to initiate chemical reactions at the arm's tip. Extremely precise chemical reactions should allow molecules to be built by the assembler. The assembler could even build additional assemblers.

Although Assistant Director Skinner has a close and nearly fatal encounter with nano assemblers in the *X-File* episode "S.R. 819," nanotechnology still exists only within the highly theoretical world of computational chemistry. Believers in nanotechnology, however, think that it is only a matter of time before such "molecular manufacturing" leaves the theoretical world of computational chemistry and enters the real world. If so, then picture a scenario where tiny molecular machines enter a person's body and repair damaged cells by reconstructing proteins, DNA, membranes . . . even making entire cells from scratch. If this is a vision of the future, then ischemia and freezing damage to cryonics patients could be repaired by completely reconstructing the brain—which would be possible if the position of every atom in the brain was known. Ralph Merkle, a proponent of cryonics and nanotechnology, believes that it should be feasible to store a digital description of every atom in the brain using computers of the future, information that could be used to repair and reanimate the brains of the frozen dead with their memories intact.

If nanotechnology becomes a reality, it will affect this planet far more than simply providing a boost for the cryonics industry. Pollution would disappear because manufacturing would no longer produce by-products. Everything would be recyclable. Diseases could be cured by little biomedical nanomachines that monitored the body for foreign invaders and destroyed cells that turned cancerous. This rosy scenario, however, may not be as utopian as it sounds. Nanomachines that can assemble anything could also disassemble everything. That's everything with a capital *E*. Proponents of nanotechnology believe that this new technology will become part of our reality in about ten to fifty years. Disbelievers think nanotechnology will go the way of cold fusion—relegated in a few years to the dustbin of interesting ideas that simply don't work.

While Arthur Grable may have bet his future on the success of nanotechnology, he won't have the opportunity to find out if he was correct. As it turns out, the killer of Arthur Grable's colleagues is Arthur Grable. Arthur's head is not pleased that his fellow scientists are carrying on his work. After all, he's not buried in the ground; he's frozen in a tank. Arthur isn't going to let a little complication like pulverization in a car accident keep him from perfecting his theory on jet propulsion, not when Roland is available to be psychically manipulated (by a dead frozen dude—only on *The X-Files*!). Roland turns out to be Arthur's identical twin brother, who was institutionalized as a child. Mulder speculates that Arthur is able to control Roland because of the connection between twins and because Arthur's frozen brain has developed psychic abilities that conscious minds are too busy to explore. Mulder's explanation is not well grounded in science. While cryopreservation is known to cause some brain cells to increase in size, the destruction of brain tissue caused by the process of freezing combined with a total lack of oxygen should pretty much eliminate all brain activity, normal or psychic.

When Arthur's remaining colleague, Dr. Nolette, realizes that Arthur is on a murder spree, Nolette pays a visit to Arthur's capsule and raises the temperature, causing Arthur's head to thaw. This eventuality is the fear of all cryonics patients. What if their cryonics company goes bankrupt? Who will keep the capsules frozen? Not all patients of cryonics companies have remained in their frozen state. In the mid-seventies, the Cryonics Society of California was freezing people for about five years before going bankrupt. With no one adding liquid nitrogen to the capsules, eight bodies thawed at the Chatsworth, California, cryotorium. Relatives of three of the deceased sued the company and were awarded almost a million dollars. Come to think of it, maybe Roland is smarter than he looks and planned the entire series of events so that a similar lawsuit could be filed on behalf of his thawed brother. It would make him a very wealthy man.

Eating Healthy

Mr. George Kearns, resident of the small town of Dudley, Arkansas, lives an uneventful life in the *X-Files* episode "Our Town." By day he is a mild-mannered USDA poultry inspector, watching an endless parade of headless chickens pass by his inspection stand at the Chaco Chicken factory. By night, he is a middle-aged Casanova, trying to recapture his youth by romancing young women. Life is good . . . especially when sweet young Paula suggests a late-night romp in the woods. However, before George can run after the playful Paula, he experiences another of his blinding headaches. Perhaps the headache is the reason why George doesn't stop to wonder why the granddaughter of the founder of Chaco Chicken is interested in someone twice her age who is also trying to close down the chicken factory, the life's blood of the town. As George runs into the woods after Paula, he discovers too late that there are worse headaches than the one he is experiencing and that chickens aren't the only creatures to face the ax in Dudley.

A federal poultry inspector missing for ten weeks doesn't sound like much of a case to Dana Scully, although Fox Mulder is intrigued by the large circles of blackened earth near the town. While Scully suggests that they are the remnants of large barbecues, Mulder attributes the scorched ground to fox fires, a local legend of pale blue balls of fire that lure unsuspecting travelers deep into woods and swamps. After the local sheriff clears up the mystery by claiming that they are merely sites of trash burning, the case seems even more mundane. Their new theory for George's disappearance? George was about to submit an uncomplimentary report to the USDA, for which he was murdered. The only question remaining is the location of the body.

It has been said that the manufacture of sausage doesn't bear close examination. The same might be said for chicken parts. Paula, who works on the chicken disassembly line, is now suffering from the same blinding headaches as George. Suddenly, Paula starts convulsing, then becomes uncharacteristically violent while wielding a sharp knife. Before Paula can harm anyone, she is shot dead by the sheriff. Her body falls into a large vat of leftover chicken parts and grain, which is normally ground together into a tasty chicken chow.

A local doctor informs Mulder and Scully that Paula was suffering from the same symptoms as George. The autopsy on Paula shows that her brain was riddled with holes, making it look like a sponge. According to Scully, Paula was suffering from a "spongy" brain disease called Creutzfeldt-Jakob. If Paula had not been felled by the sheriff's bullets, she would have died within a year of her first symptoms since there is no cure, let alone treatment, for Creutzfeldt-Jakob disease, the much-written-about human version of mad cow disease.

There are about fifteen spongy brain diseases (scientific name: spongiform encephalopathies) that can affect animals like hamsters, sheep, cattle, cats, mink, mule deer, and humans. Although George Kearns's brain wasn't available for an autopsy, he was apparently suffering from the same symptoms as Paula, and therefore probably had the same rare disease. While violent headaches are not usually a symptom of Creutzfeldt-Jakob disease, sudden personality changes, such as Paula's psychotic episode, are indicative that something is wrong with the brain. Animals who suffer from spongy brain diseases usually become lethargic and irritable. Sheep, by far the most commonly afflicted animal, lose coordination and develop such an itch that they scrape off their hair, which is how the sheep version of Creutzfeldt-Jakob disease acquired the name "scrapie."

Scully calls Creutzfeldt-Jakob disease a "prion" disease. If you think of infectious diseases as being caused by bacteria, viruses, fungi, or parasites, you may be wondering, just what is a prion? Prions are not bacteria, virus, or fungi. The disease agent that causes brains to resemble a hunk of Swiss cheese is like no other infectious agent on earth.[5]

The search for the cause of Creutzfeldt-Jakob and other similar human diseases like Gerstmann-Straussler-Scheinker disease and fatal familial insomnia was a scientific X-File that took thirty years to solve. Back in 1972, Stanley Prusiner, a neurology resident at the University of California School of Medicine at San Francisco, lost a

---

[5]If you are also wondering how to pronounce *prion*, it's "preeon."

patient to Creutzfeldt-Jakob disease and so was determined to learn more about it. In his search of the scientific literature, Prusiner discovered that the disease could be transmitted by injecting extracts of diseased brains into the brains of healthy animals. What was inexplicable was the consequence of first treating the diseased brain with radiation to destroy any DNA or RNA. Such treatment should easily kill any bacteria, virus, fungi, or parasites in the diseased brain. When treated diseased brain tissue was injected into the brains of healthy animals, they still became sick. This result was astounding because it implied that whatever was transmitting the disease had no need for DNA or RNA. Every infectious agent on earth needs to get inside another organism so that the agent can reproduce and make more of itself. Reproduction absolutely requires a genetic (DNA or RNA) blueprint. Scientists began to joke that if a virus or bacterium didn't cause Creutzfeldt-Jakob disease, the disease agent must be linoleum or kryptonite.

Prusiner, who was skeptical about the kryptonite theory, set out to find the ingredient in diseased spongy brains that was capable of transmitting the disease. He meticulously divided up the ingredients in diseased animal brains, and then patiently injected individual ingredients into normal animal brains until he found the one that could transmit the disease. His conclusions stirred up as much controversy as if he had claimed the infectious agent did indeed come from outer space. Prusiner announced to a highly skeptical world that the disease agent was a simple protein. Not a bacterium . . . not a virus . . . but just a single, simple protein. Prusiner called this new disease agent a prion.

For a protein to be a disease agent, it must be able to make more of itself, just like a virus or bacterium. But how can a protein all by itself cause even more of the same protein to be made? Proteins are always specified by DNA genes so it seemed an impossibility that a gene wasn't needed. Imagine being told that the baby you are cooing over in a carriage doesn't have a mother or father but just popped into existence one day. This is analogous to what scientists were being asked to accept. Many scientists became even more skeptical when it turned out that the genome of animals contains a gene for the prion protein and makes the very same prion protein in their own brains that Prusiner found in diseased brains. So if everyone already has that protein in the brain, the skeptics pointed out, why isn't everyone's brain riddled like Swiss cheese?

The questions were again answered by Prusiner's careful research. Yes, it was true that the prion protein causing the disease and the normal prion brain protein have the same, identical chain of amino acid beads. The key was finding that the two identical proteins fold up into different three-dimensional structures, much like two identical pieces of origami paper can be folded into two very different three-dimensional forms. In the same way, the disease prion protein and the normal prion protein can be identical—have the same order of amino acid beads—but their final folded structures can be different. When folded into the normal form, the prion protein is no threat to anyone; when folded into the disease form, it becomes a harbinger of death.

If Scully had time to analyze Paula's brain, she would have found it filled with prion proteins folded into the disease form. Somehow, the normal prion proteins in people with Creutzfeldt-Jakob disease get unfolded and then refolded into the diseased form, much like unfolding an origami swan and then refolding the paper into a frog. The next question faced by Prusiner was how do the normal proteins get unfolded and then refolded into the disease form? It turns out that all you need to do is add a little bit of disease prion protein to normal brain prion protein in a test tube, and the normal protein gets converted into the disease protein.

Think about your brain as being like a small town, say Sunnydale, California. The town is filled with normal people doing what normal people do. Then introduce a diseased person, a vampire, into Sunnydale. The vampire starts biting people and converting them into vampires. Every new vampire then gets busy converting other people into vampires. If there isn't a decent vampire slayer around, the entire town will soon be completely populated by vampires. The original vampire propagated itself, not by producing baby vampires, but by converting people who were already present into vampires.

This is how the disease prion protein propagates itself. When the disease protein contacts a normal prion protein, the normal protein unfolds and then refolds into the shape of the disease prion protein. Then the new disease protein gets busy converting other normal proteins, leading to a chain reaction that eventually results in the brain being filled with disease prion proteins. Prions are invisible to the immune system, so prion slayers don't exit. How these diseased prion proteins cause massive cell death leading to holes in the brain isn't known. What is known is that when disease prion proteins get into the brain of an animal they will kill the animal, but only after a long incubation time that can last up to thirty years.

Creutzfeldt-Jakob disease is extremely rare, affecting only about one in a million people each year. So how might George Kearns have acquired such a rare disease? There are three possibilities. Like 85 percent of the people who get Creutzfeldt-Jakob disease, the cause isn't known. Some scientists suspect that a sporadic new mutation in a single brain cell occurs in the gene that specifies the normal prion protein. In that cell, the mutant protein may more easily adopt the disease shape all by itself and then start the chain reaction that leads to infection. People can also get Creutzfeldt-Jakob disease by inheriting an already mutated gene from a parent; about 10 percent of the disease cases are attributed to inheritance. Both the sporadic and the inherited form of the disease strike people sixty years or older, so fortyish George likely acquired the disease by the third possibility: he ate a diseased animal. Perhaps, like a small number of people with Creutzfeldt-Jakob disease in rural Kentucky, George developed a taste for squirrel. Transmission of spongy brain diseases from animal to human requires eating the brain, eyeballs, or spinal fluid of the animal (the only places where the disease prion proteins accumulate in large amounts); George may have found that these parts of the squirrel taste least like chicken.

Since the incubation period for Creutzfeldt-Jakob disease can last from a few years up to thirty years, George could have consumed something as a child that killed him as an adult. He would also be at risk if he were treated with growth hormone as a child, since growth hormone used to be extracted from the pituitary glands of cadavers; if the cadaver had the disease, it was transmitted to the child. Since 1985, however, all growth hormone has been produced by genetic engineering methods so there is no longer a risk of contracting the disease from hormone treatments.

Exactly how George Kearns came down with Creutzfeldt-Jakob disease will never be known. More relevant, at least for Scully and Mulder, is the manner in which Paula got the disease. And not just Paula. Twenty-six people in Dudley had the same symptoms as Paula and George. Could the entire town have developed a taste for squirrel brains? Scully theorizes that someone disposed of George Kearns's body using the convenient chicken feed grinder in the Chaco Chicken factory; the chickens came down with the disease after eating George, and then people came down with the disease after eating the chickens who ate George. Mulder points out that people all over the country are eating the chickens and there are no signs of a massive epidemic. While Mulder is correct, there is a better reason why Scully's scenario is unlikely. Creutzfeldt-Jakob disease has a very long incubation period; there simply wasn't enough time for both chickens and people eating the chickens (and it must be the chicken brains or eyeballs) to come down with the disease. Different animal species make slightly different prion proteins and it takes longer for a dissimilar animal prion protein to nudge the human protein into the disease shape. Furthermore, there is no evidence that chickens get the disease or that it can be transmitted from birds to humans. The only explanation for how the Chaco Chicken employees got the disease so quickly is that they ate diseased prion proteins that were compatible with their own prion proteins. In other words, Paula and her friends had to have consumed a diseased human brain.

Looking for George's body, Mulder has a local river dredged. Nine skeletons are discovered, one of them indentified as George's based on an old injury. The bones all have polished ends as if they had been boiled and the skulls are missing. Mulder and Scully discover that in the past fifty years, eighty-seven people had disappeared in the neighborhood. Apparently George wasn't the only one who was tired of eating chicken.

The reason behind the unusual culinary tastes of Dudley denizens becomes more clear when Mulder and Scully determine that sweet, young Paula was actually sweet, middle-aged Paula. People who look like they are in their twenties are actually in their late forties. Mulder learns that it was Grandpa Chaco himself who found the secret for staying young when he was shot down over the South Pacific island of New Guinea in 1944. During his six-month sojourn with the Jale tribe, Mr. Chaco learned their secret of longevity: cannibalism.

Actually, cannibalism was associated with the Fore tribe of New Guinea. However, instead of imparting agelessness to the tribe, it ended up considerably shortening the life span of the women and children. The Fore, who were traditional highland dwellers, honored their dead by eating their remains, even the bodies of people who were afflicted with the "laughing death" disease known as kuru. Dozens of kuru victims had lost the ability to walk, talk, and eat, and all eventually died. Women and children were the main victims because they prepared the feast, handled the diseased organs, and ate the brains. Since death usually came many years after consuming someone who had the disease, a connection wasn't made between the feast and the disease. In the 1960s, a veterinary pathologist named William Hadlow noticed striking similarities in the brains of kuru victims and sheep who died of scrapie, the spongy brain disease. This connection made Dr. Carleton Gajdusek, who lived for many years among the Fore, start looking at environmental and culinary habits of the tribe, including their predilection for consuming the dead. Gajdusek showed that kuru could be transmitted to monkeys by injecting them with extracts of diseased brains, and the method of transmission suddenly became clear. When the Fore stopped practicing cannibalism, kuru as a disease of New Guinea all but disappeared.

While kuru is no longer a threat to people, the same cannot be said of Creutzfeldt-Jakob disease. Twenty-four people in Great Britain and one in France have come down with the disease in recent years, all traced to eating beef from diseased cows. The cows had the spongy brain disease known as bovine spongiform encephalopathy, the now infamous mad cow disease. An epidemic of the disease began in 1985 with a few sick cows in Great Britain. Apparently, cows were being fed a protein supplement of sheep, cattle, pig, and chicken parts, turning the normally vegetarian cows into cannibals. Since scrapie was widespread in sheep, cows were being fed diseased sheep prion proteins. While old methods of preparing cow chow apparently "killed" the prion proteins, new methods introduced in the 1970s allowed the prion proteins to survive long enough to begin the chain reactions in cow brains.

It wasn't until the epidemic was in full swing in 1988 that officials in Great Britain decided that, gosh, maybe cows shouldn't be eating the brains of diseased cows and sheep! By then, hundreds of cows a week were being diagnosed with mad cow disease. Because of the disease's long incubation time, the epidemic didn't peak until late 1992, when a thousand cows a week became sick. Still, no one stopped human consumers from eating the diseased beef. Two reasons were given. First, no one had become ill from eating diseased sheep so British officials hoped that cow prions also weren't transmissible to humans; and second, brains and eyeballs, the major places where the prion proteins were thought to hide, aren't usually the cuts of meat enjoyed by most people.

The hopes proved false. In 1996, doctors in Great Britain reported that ten people had died of a new variant of Creutzfeldt-Jakob disease. Instead of the usual symptoms of forgetfulness and odd behavior, victims reported feeling depressed, had trouble sleeping, and suffered from anxiety. This led physicians to refer their patients to psychologists and not neurologists. The new disease variant also took longer to kill—thirteen months from the first symptoms instead of the usual six months. But the most disturbing aspect of the new variant was the age of the victims. Instead of people sixty years or older coming down with Creutzfeldt-Jakob disease, the average age was twenty-seven. It soon became clear that these unfortunate Britons were the first to have acquired the disease from eating sick cows.

Because the cow disease peaked in the early 1990s, and because of the long incubation time for traditional Creutzfeldt-Jakob disease, scientists are wondering if these will be the only victims—or just the tip of an epidemic iceberg. Since the disease prions can be detected in tonsils and the appendix, the British government is embarking on a study to test if such tissues that have been removed from thousands of people and stored contain any of the disease proteins. If only one of the one thousand tissues to be tested is infected, this would translate into fifty thousand cases. Whether such an individual would be told the dismal news is an ethical dilemma that the British have not yet solved. Regardless of the outcome of the testing, the crisis is unlikely to extend to other countries; while 170,000 cases of mad cow disease have been found in Great Britain, only a few hundred sick cows have been found in other European countries and never yet in the United States.

The cannibalism practiced by the citizens of Dudley only leads to eternal youth on *The X-Files*. Not that people throughout history haven't tried equally bizarre methods to slow the aging process. French physiologist Charles-Edouard Brown-Séquard told people in 1889 that rejuvenation could be achieved by inoculating one's arms with the crushed extracts of domestic animal testicles. This idea did not amuse the medical journals—to say nothing of the animals—but led many others to experiment with other types of testicular rejuvenation therapy. In the 1920s, John Brinkley, a fundamentalist preacher, grafted goat testicles onto thousands of old people, since goats were such sexual athletes. Also, in the 1920s, doctors at San Quentin prison in California implanted testicles of executed felons into other prisoners to determine if it increased virility and slowed the aging process. Such treatments, like any surgical grafts, must have led to massive immune responses and rejection of the implanted tissue, which could not have enhanced anyone's longevity. With all that is known about the immune system these days, it is astonishing that some people are still injecting themselves with tissues of dead animals in the mistaken belief that it will slow down aging.

Scully, it turns out, was right about the large circles of blackened earth on the outskirts of Dudley. They are indeed the remains of large barbecues. At the end of the episode, the whole town of Dudley turns out for an evening's menu that features Extra-Crispy Grandpa Chaco and Scully soufflé. Fortunately, Mulder arrives on the scene in the nick of time, spoiling everyone's appetite. In the end, the Chaco Chicken factory is closed down. No longer will the townsfolk live by the Chaco Chicken motto: "Good People, Good Food."

6
Fooling with Mother Nature

Introduction

INT. MODEL HOUSE—NIGHT

MULDER
I'm in a house. It's apparently empty.

A MOTORIZED NOISE is heard. Mulder flashes his light upwards. In the corner of the ceiling, a video camera is shifting its position until it's pointed at Mulder.

MULDER
Except for the camera on the ceiling.

SCULLY
What's the place look like?

MULDER
It's just a normal, suburban, two-story house. A large living room, nice carpeting, sparsely furnished, fireplace. A kitchen, wallpapered.

SCULLY

It sounds like you're describing my dream house.

MULDER

Except for the moving walls.

SCULLY

The walls are moving?

—"War of the Coprophages"

Pop quiz. Imagine that your house is your entire world. The bathroom supplies water and one bedroom is converted into farmland. The kitchen is your manufacturing base. Beyond the translucent outer walls hangs a strong source of light and radiation. There is no outside trash dump, so any waste that isn't naturally consumed by microbial housemates gets thrown into the basement—which is nearly full. Smoke and chemicals from the kitchen can't be vented "outside" so they gradually permeate the entire house, weakening the walls and letting in damaging radiation. Artificial chemicals cooked up in the kitchen but never produced by nature cannot be broken down by nature's organisms. These chemicals end up everywhere—on walls and floors, in the water supply, and on crops—and eventually inside all the house's inhabitants. Species that have been beneficial, unobtrusive housemates for years start having babies that are too feeble to survive and, one by one, these species disappear. Given this scenario, do you

(A) Continue as if nothing is wrong
(B) Declare the kitchen off limits and sit in a dark bathroom gnawing on raw broccoli
(C) Press the reset button
(D) Pay closer attention to what's causing the problems and start solving them

For the past two hundred years, the majority of human residents on the planet have answered (A). A few have answered (B) but it hasn't been a popular choice. Picking (C) requires finding the button, but pushing it doesn't guarantee better execution the next time around. (D) isn't a cheap or easy answer. Manufacturers didn't pollute the planet with PCBs and dioxins as a way to kill off bald eagles and whooping cranes. CFCs weren't used as refrigerants in order to get rid of that ozone layer. Machines that burn fossil fuels weren't designed to produce smog and toastier temperatures. Decisions were made and are still being made that reflect the bottom line and not some uncertain future harm. And what if you pick the expensive answer (D) but (pardon the analogy breakdown) neighbors adjoined on all sides can only afford (A)? This is a major concern since decisions made by one house affect every house.

Environmental issues are like fires raging out of control in remote regions far from where people are trying to live in relative peace. Occasionally a hint of smoke is in the air, but the fires have only a minor impact on the lives of most people. Putting out the fires requires much expense and possible injuries. Without intervention, the fires might burn themselves out without harming any civilized areas, so any money spent to fight the fires might be considered wasteful. But the fires could grow and swing into populated areas. No one would argue to leave these fires alone, but fighting them now is far more costly. Whether containing the current environmental fires will be worth the sacrifices and money spent today may never be known. Science unfortunately cannot predict with absolute certainty that elimination of the ozone layer, global warming, and widespread species extinctions are the inevitable outcomes of our current industrial societies. Governments therefore will continue to grapple with the issue of whether higher costs, escalating taxes, and lower standards of living are necessary to preclude scenarios from occurring that might never happen. Fires in remote regions are often allowed to burn themselves out for many reasons. But no one proposes allowing them to burn unhindered by any controls, just like it seems a very foolish risk to continue answering (A) when science is telling us that so much is going wrong.

As X-Files fans know, many series episodes have woven environmental messages within the main plot lines. Transplanted cockroaches terrorizing a small town . . . missing links emerging from a shrinking habitat . . . mutants arising from global warming—these are just some of the messages that speak to real environmental concerns.

"Transplanted cockroaches terrorizing a small town" is the theme of the *X-Files* episode "War of the Coprophages." Before Fox Mulder discovers that the six-legged invaders of Millers Falls are actually little metal cockroach machines beyond the manufacturing capability of modern science, Dana Scully comes up with a logical explanation for their presence and strange attraction to humans. She believes that the cockroaches are an exotic species, transplanted deliberately or inadvertently to the Millers Falls area. And just like other exotic species that suddenly find themselves in a world without traditional enemies, the foreign cockroaches are free to grow unrestrictedly.

Every species on Earth has a home where it originated and subsequently evolved along with neighboring organisms. Individual species populations are kept in check because pathogens, parasites, and predators also evolved in the same locations. As a scientist, Scully would be aware that when a species is removed from its natural ecosystem and introduced to a new neighborhood where enemies are few, the results can be devastating for native plants and animals, or in the case of Millers Falls, resident humans. Consider what happened in Australia with the arrival of British settlers. Civilized society meant hunting, so wealthy Englishmen brought along a few hundred rabbits for a bit of sport. The rabbits found themselves in paradise. Before long, hundreds of millions of rabbits occupied the country, destroying natural vegetation and crops and causing severe soil erosion. In the late 1970s, the government let loose a virus that wiped out 99 percent of the rabbits. The remaining rabbits were resistant and the population soon recovered. Then ferrets were imported to kill the rabbits. But the ferrets were infected with bacteria that causes bovine tuberculosis, which spread to cattle and deer. So another virus was brought in to kill the ferrets. Recently, the government started testing one more rabbit-killing virus called a calcivirus on an island off the Australian coast. Before testing was completed, the virus escaped and is now spreading through Southern Australia, killing 90 percent of adult rabbits. Scientists are currently keeping fingers crossed that rabbits are the only hosts for the incompletely tested calcivirus.

There are far too many other examples of misguided introductions of exotic species to mention them all here. In the United States alone, four thousand species of exotic plants flourish outside cultivation, and four hundred are growing so aggressively that they are wiping out native plants and animals. Kudzu was introduced as a beautiful, fast-growing ornamental vine in America in the nineteenth century. In the 1930s, the U.S. Department of Agriculture even paid farmers to plant the vine to help control erosion throughout the South. The vines have since taken over huge areas, covering houses and hillsides, killing native vegetation and crowding out animals. Each vine spreads over sixty feet a year and can take ten years to kill. Melaleuca, or paperbark trees, were also planted for what seemed like a good reason—drying swampland in Florida. One tree can become six hundred feet of dense canopy in one year, not a desirable outcome anymore given the importance of maintaining the Florida Everglades and its unique species. As a result, the melaleuca trees are now cast in the role of villains that threaten the entire region. To slow down the invasion, the U.S. Department of Agriculture is releasing another exotic species—Australian snout beetles—which should feed on melaleuca trees and nothing else . . . hopefully. Another misguided introduction of exotic species occurred in 1890. European starlings, which granted don't have quite the same effect on one's sensibilities as exotic cockroaches, were released in New York's Central Park as part of a lofty plan to import all birds mentioned in Shakespearean literature. Starlings are now the most abundant bird in North America and cause widespread damage wherever they flock.

Sometimes, the introduction of an exotic species is unintended. On the island of Guam, brown snake stowaways on cargo ships have caused the extinction of all native songbirds including twelve species found nowhere else. In 1860, a French artist trying to breed a better silkworm inadvertently let loose a French gypsy moth in Medford, Massachusetts. It took only ten years for the moths to defoliate every tree in the neighborhood and today, the moths are a major pest in the Northeast. Some years, so many gypsy moth caterpillars fall from leafless trees that it sounds like noisy rain falling from the sky.[1]

[1]Sometimes it isn't the species itself but what it harbors that becomes an ecological nightmare. Around the turn of the twentieth century, healthy-looking chestnut trees were imported from Asia and sold for the reasonable price of 25 cents to one dollar. Only after the trees were planted near native American chestnuts in New York and Connecticut was it discovered that the Asian trees harbored a fungus called chestnut blight. The Asian trees had evolved alongside the chestnut blight fungus and thus were accustomed to its presence and

"Missing links emerging from a shrinking habitat" will be familiar to *X-Files* fans as a description of the early episode "The Jersey Devil." The environmental message in "The Jersey Devil" is very clear: destruction of natural habitats leads to species declines and extinctions—in this case, a line of feral humans.

The loss of species from this planet because of habitat destruction is occurring at an alarming pace. Consider only vertebrate animals—animals that possess an internal skeleton and a spinal column. Two thirds of all bird species are in decline, with 11 percent in danger of extinction. Twenty-five percent of all mammalian species are in trouble, with nearly half of our nearest relatives, primates, threatened with extinction. Twenty to 25 percent of reptiles and amphibians are declining in numbers. But the worst problems are faced by fish. Estimates are that 13 percent of all fish species are in immediate danger of extinction with another 21 percent vulnerable to extinction. The forty thousand dams blocking the world's rivers, the runoffs from pesticide-treated crops, and the destruction of coral reefs mean the end is coming for over seven hundred species of fish.

Habitat destruction in the tropical rain forests is a significant contributor to loss of global biodiversity. Although tropical rain forests cover less than 5 percent of the Earth's surface, they are home to nearly half of the world's species. North America contains only four hundred species of trees and 750 different butterflies. A single four-square-mile patch of rain forest is home to some 750 different species of trees and well over a thousand different kinds of butterflies. One percent of the rain forests are lost to logging and field clearing each year. Sometime in the next century, the rain forests with their millions of trees, mammals, birds, and insects—some found nowhere else in the world—may be reduced to a memory.

The fossil record suggests a background rate of extinction of about one to three species a year. Today the estimate is one thousand

---

remained outwardly healthy. The American chestnuts had no such advantage and the fungus quickly found millions of defenseless chestnut trees ripe for infection. By 1908, the fungus had spread into Pennsylvania, Ohio, Virginia, and Massachusetts. The mighty American chestnut trees, once so common in forests and backyards throughout the Northeast, are now reduced to ugly stunted shrubs.

species a year, a number not seen since the extinction of the dinosaurs. Our descendants won't be combing the oceans for some catastrophic meteor crater as the reason for the disappearance of so many of the Earth's species during the twentieth and twenty-first centuries. They will be pointing their fingers at us.

"Mutants arising from global warming" . . . having trouble coming up with the episode to match this phrase? That's because it isn't an X-Files. By the time you finish reading below, you'll wish that's all it was.

Depending on who you talk to, global warming is either an unproven possibility or the inevitable outcome of an industrial world. According to many (but not all) environmental scientists, global warming, also known as the greenhouse effect, is what awaits a planet that spews billions of tons of carbon dioxide into the atmosphere from the burning of fossil fuels while at the same time destroys the rain forest trees that help soak up the problem. Whether or not you believe that dumping carbon in the atmosphere and global warming are linked, it cannot be denied that carbon emissions are increasing (from 1.5 billion tons in 1950 to almost 7 billion tons today) at the same time that the world is experiencing the hottest temperatures on record. The warming trend could be an anomaly, a transient fluctuation coincidental to the release of greenhouse gases. Unfortunately, climate cannot be reproduced in a laboratory experiment. The experiment must therefore occur all around us and the results should become very clear in the next century. But the damage may be difficult to undo at that late date.[2]

In 1995, the Intergovernmental Panel on Climate Change estimated that the next century will see a doubling of carbon dioxide emissions in the atmosphere and an increase in global temperature of

[2]A correlation has been found between local pollution and local weather patterns. For those people living in metropolitan areas in the Northeastern United States, you're not imagining that rain often seems to wait for the weekend. Climatologists from Arizona State University have found that pollution levels increase steadily during the work week—lowest on Mondays and highest on Thursdays, remaining nearly as high on Friday. The rain cycle is also weekly, with Saturday and Sunday getting the most rain. Since the seven-day week was not concocted by Mother Nature, the unusual seven-day rain cycle is thought to be directly related to cycling pollution levels.

2.2 to 6.3 degrees F—the fastest change in the past ten thousand years. If this scenario takes place, global warming and other environmental stresses could spell rough times ahead for the creatures of this planet. Besides the more traditional predictions of coastal flooding due to melting arctic ice and sweltering summer temperatures, a new and very unexpected concern has recently emerged. In an article published in the journal *Nature*, University of Chicago scientist Susan Lindquist and her colleagues found that environmental stresses like hot temperatures can cause substantial changes in nature's creatures within a single generation. Using fruit flies, Lindquist discovered that high temperatures caused normal-looking flies to have babies with all sorts of deformities. Some baby flies had rough, smooth, or missing eyes; others had strange-looking legs or wings. The deformities were passed on to the next generation, indicating that the DNA itself was mutated. The stress itself wasn't causing the mutations. The mutations were already in the population of flies, but something was masking the mutations—covering them up so that the flies continued to look normal. Normal, that is, until an environmental calamity like sweltering temperatures caused the mask to drop off and the mutations to show their true stripes.

The mask on the mutations is a protein called Hsp90. Hsp90 plays a role like the police in a large urban city that keep a criminal element under control. The criminals are present, but the police keep an eye on them and the city as a whole is not concerned. If a calamity in one neighborhood causes all the police to be pulled off their normal patrols, the criminal element is now free and becomes highly visible. The resulting crime spree changes the entire character of the city, which may never revert to normal even when the wayward police come back.

This is what happens with Hsp90. The Hsp90 protein functions like the police, keeping mutant criminal proteins from having any effect on the organism. But Hsp90 has another role to play when the organism is severely stressed. Hsp90 stops controlling the mutant criminals and instead is co-opted into helping the organism survive. Without Hsp90 to mask the mutant proteins, the defective proteins are free to affect the development of the stressed flies' babies, which are born with physical and possibly mental deformities.

Why, you might ask, did organisms develop such a system,

where all sorts of mutants come to light when the population is stressed? One possibility is that it allows rapid evolution during times when an organism needs it most. Lots of babies would be born with all sorts of physical differences, some of which might be beneficial for surviving the stressful conditions. Of course, most of the physical deformities would be highly detrimental. Humans also have the Hsp90 protein . . . and global warming could be a severe environmental stress. No, "mutants arising from global warming" isn't an *X-Files*. It's from the program called *Real Life* and we're all starring in it.

The *X-Files* does cover some additional environmental themes. Chemicals cooked up in the industrial kitchen in the episode "Blood" affect more than just their target insects; biological warfare as depicted in "The Pine Bluff Variant" is the ultimate poisoning of the environment; and disruptions in the food chain caused by diminishing amphibian populations lead to a disturbing substitute in the episode "Quagmire." These *X-Files* remind us that nature is not just a picture postcard. She is a living system. And we are fooling ourselves if we believe that we can continue to fool with Mother Nature without someday paying an enormous price.

Secret Scents

INT. ELEVATOR—DAY

TABER'S POV—ELEVATOR DIGITAL DISPLAY
"NO AIR"
TABER
His phobic reaction intensifies. His eyes are locked upon . . .
TABER'S POV—ELEVATOR DIGITAL DISPLAY
"CAN'T BREATHE."
TABER
GASPS for air.
ELEVATOR
The cramped passengers tense. A few turn and look back at the obviously uncomfortable man.
TABER
In a cold sweat.
He trembles, eyes glazed, homicidal.

TABER'S POV—DIGITAL DISPLAY
"KILL 'EM ALL."

—"Blood"

In the *X-Files* episode "Blood," all is not well in Franklin, Pennsylvania. In the past six months, Franklin has undergone a metamorphosis from sleepy hamlet into the murder capital of America. The blame cannot be placed on drug dealers, gang members, or even invading Martians. The killers are ordinary citizens like housewives and real estate agents. The latest is a man suffering a bout of claustrophobia in an elevator. To relieve the tension on the way down, he decides to murder his fellow passengers with his bare hands.

Fox Mulder is at a loss to explain the recent murder sprees when he is called in to help an overwhelmed police force. The only connection between the various killing rampages is the destruction of electrical equipment at the crime sites and the deaths of the suspects when they refuse to give up. Mulder finds only one other clue at the latest bloody scene—a seemingly harmless chemical residue beneath the nails of the dead perpetrator. Not even Mulder's all-purpose fallback theory—paranoia due to mass abductions by aliens—fits the relevant data. When Mulder nearly becomes murder victim number twenty-four by asking innocent questions of a woman in her home, it is clear that something has gone very wrong in the town of Franklin.

If residents of a city have an abnormal rate of cancer incidences, scientists called epidemiologists search for a possible environmental factor that is putting the citizens at risk. Likewise, Mulder feels that exposure to something in the environment is turning selected Franklin residents into murderers. He begins to make headway in his investigation during a morning jog when he notices a truck dumping something into the front yards of homes along a residential street. The dumped material isn't a toxic chemical, but rather thousands of flies. When Mulder shows a fly to his longtime friends—the science savvy, conspiracy-buff trio called the Lone Gunman—he learns that it is a Eurasian cluster fly, an annoying pest of homes, schools, hospitals, and other buildings in the U.S., Canada, and Europe. In the summer, cluster flies ravage flowers and crops; in the winter, they escape the cold by sneaking into buildings through minuscule cracks and hiding in attics or between walls. An infestation of crawling cluster flies can make ceilings of homes appear black. Plastering insecticide only succeeds in transferring the problem since the evicted cluster flies will scout long distances for other suitable lodgings.

Instead of the normal (and not very effective) remedy for cluster fly invasions—sucking them into a vacuum cleaner—Franklin officials decide to dump large numbers of additional flies into infested areas, as witnessed by Mulder. While adding more flies may not seem like the best choice for ridding the pests from neighborhoods, Mulder determines that the dumped flies have been sterilized by being purposefully bombarded with radiation—"nuked," as described by Mulder's Lone Gunman friend Frohike. The principle behind dumping nuked flies is simple: Introduce massive numbers of sterile male insects into a problem area and the sterile males will mate with females without any baby insects being born. The population size goes down without the release of harmful chemicals. While this method makes simple intuitive sense, nature is never simple.

Controlling insects using a sterile insect program like the one being tried in Franklin requires the ability to grow millions if not billions of insects in a "factory," to separate the males from the females (imagine what a delightful job that must be), and then to sterilize the males before releasing them into the community. While it may seem that growing insects in a factory should be simple—after all, they do quite well in one's own home—rearing millions of robust insects is a daunting task. The insects grown in a factory must be Adonises of their species to attract discriminating females when released. Unfortunately, this is rarely the case. Despite a great deal of expense and tender, loving care, insects reared under artificial conditions often have behavioral and physiological problems that cause wild females to turn up their noses. One has only to eye the pathetic-looking creatures released by the town of Franklin to see the problem; what female could possibly want these poor excuses for cluster flies as they stagger about staring in envy at their virile wild brothers? Since factory-bred male insects rarely measure up to wild males, an overwhelming number are needed to make a dent in the pest population—as many as one hundred sterile males for every one wild male.[3]

[3]Sterilizing the males is also tricky. The process must not further detract from their already suspect ability to impress the ladies. Since little fly vasectomies

The town of Franklin has the best of intentions in initiating a sterile fly program to control their cluster fly problem. They are not releasing exotic species or dumping harmful chemicals into their community; rather, they are taking an environmentally friendly approach. Before beginning the program, town agricultural officials probably hoped that they would have the same success as farm communities in the 1960s that released billions of sterile screwworm flies. The sterile screwworm fly program completely eliminated this agricultural pest from the U.S. without the need for chemical insecticides. Unfortunately, other sterile insect releases haven't been anywhere near as successful—as Franklin officials discover when their cluster fly problem doesn't go away.

When Mulder determines that the town has a cluster fly problem that is not being solved by the release of sterile insects, he surmises that other measures are being taken. Precisely what measures become clear when Mulder's Lone Gunman friends discover that the substance found under the nails of the killer is an experimental synthetic botanical insecticide. Botanical insecticides are extracts from plants that are toxic to insects and other animals. The Lone Gunman are not fans of this particular botanical insecticide and equate it with the infamous chemical insecticide DDT, once widely used and now banned.

What the Lone Gunman fail to mention is that botanical herbicides are generally considered to be safer for the environment than synthetic chemical pesticides like DDT. One reason is that natural plant products are quickly broken down into simpler, less toxic substances. Compounds like DDT that were never formed in nature can persist in the soil for fifty years or longer. Many such chemicals also survive inside animals unfortunate enough to eat the sprayed plants, and animals who eat those animals, and insects that eat the remains of animals who ate the animals, and fish that eat the insects, and so

---

are impractical, the preferred method of sterilization is radiation treatment of juvenile insects just before they produce sperm. If the correct amount of radiation is used, the flies themselves will suffer few abnormalities but the DNA in their sperm will have large numbers of mutant genes, some of which will be lethal to offspring that develop from eggs fertilized with these sperm. If the radiation procedure falls short of complete sterilization, large numbers of still fertile flies will be released, exacerbating an already severe problem.

on up and down the food chain. The synthetic chemicals are also passed down to subsequent generations. So when synthetic chemical insecticides are determined to be more harmful than previously thought and banned from further use, what was already used remains with us for a very long time.

Byers—the most erudite member of the Lone Gunman—who never met a conspiracy theory that he didn't think of first, tells Mulder that DDT was shown to cause higher rates of breast cancer in women. Although Byers makes this statement as if it were fact, such conclusions about DDT are still enormously controversial. In making such a statement, Byers may be referring to reports that link a one percent rise in the incidence of breast cancer since the 1940s with the increasing use of DDT and other chemicals that fall into a class of agents called environmental estrogens.

Estrogen is a natural hormone that is one of nature's ingredients for making women different from men. Estrogen is a member of a class of hormones called steroid hormones, which can easily slip through a cell's outer membrane. Some specialized cells make a protein called the estrogen receptor that floats around in the nucleus of the cell. After estrogen enters a cell, it slips into the nucleus through the membrane surrounding the nucleus and then "fits" together with its receptor like two puzzle pieces. The unit formed is then able to attach itself to DNA and cause certain genes to be turned on and off. The production and release of hormones like estrogen in an organism is tightly regulated because even minute amounts can have far-reaching effects on reproduction, behavior, and development of the brain.

Environmental estrogens are compounds that mimic estrogen. Even though environmental estrogens don't look like natural estrogen, they can still get into cells and fit together with the estrogen receptor, which is then fooled into interacting with DNA. Environmental estrogens can be man-made chemicals that don't normally exist in nature, like DDT, or they can be compounds found in many plants—the so-called phytoestrogens. Large doses of these environmental estrogens cause abnormalities in many animals. Consider, for instance, the fate of alligators living in Florida's Lake Apopka. Following a massive spill of a DDT-like pesticide in the early 1980s, the male alligators suffered problems in the development of their masculine traits. In addition, problems with infertility and timing of sexual maturation plague both fish living downstream of paper mills that spew high levels of plant phytohormones into the water and cows grazing on phytohormone-rich plants.

There is no question that a woman's lifetime exposure to natural estrogen is one factor affecting the risk of contracting breast cancer. What is much more controversial is whether, as Byers believes, low-level exposure to environmental estrogens like DDT also contributes to cancer. While plant phytoestrogens seem to have a protective effect against breast cancer, the possibility that synthetic environmental estrogens have the opposite impact is currently the subject of a substantial debate. In one early study, scientists examined women who had breast cancer and those who didn't for levels of DDT and other environmental estrogens. Although this study showed that women with breast cancer had higher levels of environmental estrogens in their systems, larger, more recent studies show no such link. What is exasperating scientists who are trying to set up logical experiments to tackle this extremely important issue is that there are no women anywhere on the planet who don't have environmental estrogens in their bodies (men too). Without such a "control" population group, it's impossible to determine a baseline risk for breast cancer in the absence of exposure to environmental estrogens.

Another problem with these types of studies is that unlike those using genetically identical laboratory animals that can be kept under similar stress levels and fed a strict diet, there would be few volunteers for scientific studies if it meant living in cages and dining on pellets. People eat different foods, live in different places, endure different levels of stress, and are exposed to different chemicals in the environment. What many scientists fear is that low doses of environmental estrogens on top of other genetic or environmental factors can cause cancer or other abnormalities in a subset of the population. It's these additional kinds of risk factors that contribute to the problem facing the citizens of Franklin, Pennsylvania. Only a few people became killers, so whatever agent is causing this transformation isn't having the same effect on everyone.

Regardless of whether environmental estrogens are responsible for the increase in breast cancer cases, they still comprise some of the nastiest members of the chemical Hall of Shame: DDT, the insecticide that nearly wiped out eagles and other bird populations by thinning the shells around eggs; DES, given to pregnant women to prevent miscarriages until their children were discovered to have all sorts of abnormalities, including rare cancers; and kepone, an insecticide used in the 1970s that afflicted employees in facilities where it was manufactured with tremors, chest pains, and other problems.

Man-made chemicals don't have to mimic estrogen to have an adverse affect on people. A growing number of artificially manufactured substances are now known to lower IQ, produce antisocial behavior, affect motor coordination, and cause many other physical and neurological abnormalities. Consider PCBs, used in the construction of electrical transformers and capacitors, and dioxins, produced as a by-product in the manufacture of some pesticides. When inside a person or other animal, PCBs and dioxin block the action of some necessary hormones and also cause the liver to produce enzymes that help eliminate hormones from the bloodstream. When these important hormones, such as those made by the thyroid, are not present in their normal amounts, mental retardation, learning disabilities, and attention deficit disorder can result. Many scientists suspect that the increasing number of children with learning disabilities and behavioral problems is not a statistical artifact stemming from better testing methods, but comes from exposure of their mothers to chemicals like PCBs and dioxins. Since environmental estrogens, PCBs, and dioxins are found in every corner of the planet and in every living creature, and are so long lived that they are passed from generation to generation, the possibility for affecting whole populations of organisms including people is a frightening scenario only beginning to receive government attention.

The Lone Gunman determine that the botanical insecticide found on the killer isn't an environmental estrogen like DDT but rather a substance that acts like an alarm pheromone on cluster flies. Pheromones are chemical odors given off by living organisms that send messages to members of the same species. The type of pheromones pertinent to the episode "Blood" are alarm pheromones, conveyers of the message "Fly for the hills, we're under attack!" Other pheromones produce a different reaction. Some female insects release sex pheromones, which spread the message "come and get me, boys" to any males within snuffing distance. There are also pheromones called aggregation pheromones that issue invitations to join in creating big communes.

Alarm pheromones, like the synthetic botanical one that interests Mulder, are not widely used as insecticides.[4] The main type of pheromone used as an insecticide is the sex pheromone. When sex pheromones are released into the air above crops, male insects become confused and can't find the pheromone traces left by their females. Sex pheromones are also used as bait for traps: the male insects are attracted to the traps by the pheromone, but instead of finding ripe young females, they meet poison and die. Since natural pheromones can only be sensed by members of a single species—the same species that released the pheromone—pheromone insecticides are considered to be an environmentally friendly way to combat problem insects.

Mulder learns from the Lone Gunman that this particular botanical insecticide has not yet been approved by the Environmental Protection Agency. To get evidence that the town is illegally spraying the insecticide on its crops, Mulder spends the night in a nearby agricultural field. During the night, Mulder's suspicions are answered when he is drenched with a chemical sprayed from a helicopter. Analysis of the sprayed chemical indicates that it is the same as that found beneath the nails of the killer. Mulder reasons that a plant compound that just coincidentally functions as an insect alarm pheromone might not be species-specific. He goes on to speculate that the botanical alarm pheromone being sprayed on the Franklin crops is alarming more than just the insects.

Insects are not alone in having the ability to communicate by pheromones. Simple organisms like yeast need sex pheromones to help them distinguish yeast of the opposite sex.[5] Plants also produce scents when they are under attack to warn other nearby plants. Since plants on the receiving end of the messages don't have the option of running away and aren't sure what's causing the problem—pheromonic messages can be scanty on the details—they make a battery of new compounds that are toxic to insects, viruses, fungi, and bacteria.

---

[4]There is one alarm pheromone that has been used to control aphids by making the insects more agitated so that they pick up more of a second insecticide sprayed on the same plants.

[5]Yes, fungi like yeast have sexes. They are called mating types.

Mammals also communicate with pheromones. Hamsters use pheromones to select their mates and recognize others in their social group. Mice sense "bad vibe" pheromones if the previous occupant of their space was stressed. For humans, though, communication is clearly dominated by what's seen and heard and pheromones aren't obviously part of the communications package. Mammals that communicate by pheromones have a tubular-like structure close to their nasal cavity called a vomeronasal organ, which is used to sense the odors that send subconscious messages to the brain. The vomeronasal organ in humans, though, was thought to be a relic that no longer functioned—no vomeronasal organ, no communication by pheromones. As it turned out, scientists were looking for the organ in the wrong place. In 1994, a functional vomeronasal organ was found near the base of the nasal septum, the structure in the nose that divides it into two equal parts. The discovery of the organ didn't, of course, tell what pheromones it sensed or what those pheromones might be making us do subconsciously.

In March 1998, the first human pheromone was discovered. It wasn't a sex pheromone to help tell apart males from females or even an alarm pheromone. It was a scent given off by the armpits of women that causes the menstrual cycles of other women in the vicinity to synchronize.[6] Why such a pheromone exists remains a mystery. As for other human pheromones, it's probably only a matter of time before scientists find that one's mood, attraction to others, and maybe even intuition and empathy are all feelings subtly dictated by subconsciously sensing the odors of people around you.

Mulder believes that the synthetic alarm pheromone is affecting people as well as insects. Since this is no natural alarm pheromone that would be specific to only one insect species, it is not completely beyond the realm of extremely remote possibility for an artificial pheromone to affect both insects and people. Mulder learns that each of the mild-mannered citizens-turned-killers lived in a heavily sprayed area. Since Mulder was doused with the same

---

[6]You can imagine how much fun it must have been to participate in this seminal study. Volunteers had secretions from under the armpits of women wiped beneath their noses with orders not to wash for six hours.

compound, Scully is called in to perform a medical examination to determine if Mulder has also become a walking time bomb.

In her prior autopsy of one of the murderers, Scully determined that the amount of adrenaline in the person's system was one hundred times higher than it should have been even for someone who suffered a violent death. Adrenaline, a hormone secreted into the body by the adrenal gland, primes the body for a "fight-or-flight" reaction. A rush of adrenaline flooding the body is something that should be familiar to anyone who has cruised at eighty-five miles per hour on the freeway and seen red lights flashing on top of a car in the rearview mirror. Imagine if that feeling of pounding heart, rising blood pressure, and quicker breathing associated with a normal release of adrenaline was amplified a hundredfold. This state of extreme panic is what some residents of Franklin must have felt by having such levels of adrenaline in their systems. Scully finds no signs of increased adrenaline in Mulder's body and concludes that the pheromone being sprayed on the plants was not responsible for the adrenaline surges in the killers and their associated psychotic behavior.[7]

Mulder suspects that the killers saw violent messages in LED displays on various pieces of equipment (hence the smashed machinery at each of the crime scenes). Since there was no obvious electrical connection between a postal sorting machine, elevator floor indicator, bank automatic teller, automobile diagnostic machine, and microwave oven, the killers had to be suffering from hallucinations as well as adrenaline overload. When Mulder is told by the Lone Gunman that the chemical composition of the botanical insecticide, lysergic dimethryn, resembles another potent plant product, lysergic acid diethylamide, or LSD, another piece in the puzzle falls into place. Mulder adds the remaining pieces when he discovers that all the killers suffered from some type of phobia—blood, fear of rape, claustrophobia—that helped trigger the violent attacks. Somehow, a combination of risk factors such as having a severe phobia, high levels of adrenaline, and

---

[7]Mulder also mentions that he was burned by the pesticide spray. Scientists have shown that such burning sensations and other symptoms like headaches, nausea, and chest tightness aren't due to the pesticide itself. Rather, the emulsifiers and propellants that allow the insecticide to be sprayed in one's house or onto crops are frequently the real culprits.

the alarm pheromone combined in an unpredictable way to create killers out of ordinary (but probably normally jittery) people. Scully suspects that some compound induced by adrenaline in the body causes lysergic dimethryn to convert into LSD. Mulder isn't affected because his adrenaline levels are normal. The killers, therefore, were under the influence of LSD when obeying messages that they saw in the LED displays—messages like "Kill 'em all." By the end of the episode, the town of Franklin has identified everyone who might be sensitive to the insecticide, stopping one individual just as he startes shooting people from atop a clock tower at the local college. With twenty-three deaths over a six-month period, Franklin officials and the viewers certainly learned a lesson about using unapproved pesticides.

But what would have kept this pesticide from being approved? The chemical was harmless on nearly everyone. The side effect of turning some individuals into hallucinating killers might not have been discovered if the compound was tested only on lab animals or healthy adult volunteers.

This is the major problem facing government agencies charged with determining the safety of any substance—from natural plant product to artificial chemicals—dumped at high levels into the environment. Scientists are now finding that sometimes only selected people in a population react adversely to an introduced compound. If not men, then women; if not women, then the fetuses inside pregnant women; if not fetuses, then children or teenagers; if not teenagers, then old people; or maybe only people who are stressed or taking a particular medication, or exposed to more than a single chemical; or maybe only people who have phobias and high levels of adrenaline. With the continuing search for more effective pesticides and the increased manufacturing of synthetic chemical compounds, this is a problem that seems destined to be with us for quite some time.

The Ultimate Insult

SCULLY'S POV—THE MONITOR

The field agent and two other agents stand before Goatee Man's prone body. We can't see the man, but we can hear his CRIES of pain over the parabolic mic.

                    FIELD AGENT
                     (filtered)
Oh my god ... what'd he do to him?

                    SKINNER
What is it? Talk to me.

                    SCULLY
I can't see anything.

THE FIELD AGENT
Looks down in horror at:
HIS POV—GOATEE MAN
Mercifully unconscious as the flesh on his hands and face is being slowly
eaten away—as if acid were spreading over him.

                    FIELD AGENT
It's ... eating away his skin.

                    SCULLY
What?

                              —"The Pine Bluff Variant"

   In "The Pine Bluff Variant," Fox Mulder and Dana Scully take a
respite from UFOs, time rifts, and alien autopsies. Their current mission: participate with ten other FBI agents in a classic stakeout. The
bait is an arms merchant sitting on a park bench. The man they are
hoping to hook is a member of a domestic terrorist group. Vigilance
soon pays off when the terrorist steps off a bus and hands an enve-
lope to the arms merchant. The arms merchant soon realizes that he
has received more then he bargained for when pain shoots through
his body and hideous scabs break out on his hands and face. As the
man lies dying, Mulder rushes after the terrorist. Scully, unknown to
Mulder, follows in close pursuit. A few minutes later, Scully is
shocked to see Mulder standing by nonchalantly while the terrorist
drives away.
   Scully and a strangely evasive Mulder return to find that the

arms merchant has died. Scully suspects that the man was killed by being deliberately infected with a toxic organism. The terrorist helped by Mulder appears to be engaging in biological warfare—the deliberate release of deadly organisms into the environment. A terrorist organization would develop and use biological warfare agents for two reasons: to kill in the most painful and inhumane way imaginable and to spread panic throughout a fearful population.

The deliberate release of disease agents isn't new to the twentieth century. History is filled with episodes where combatants tried to gain an advantage by demoralizing the enemy with the threat of infectious disease. In ancient times, rotting corpses of animals were thrown into wells to poison the water supply. By the Middle Ages, soldiers were catapulting corpses of smallpox and plague victims over castle walls to infect and panic their opponents. The first use of biological warfare in America was the well-known story of Lord Jeffrey Amherst, who in 1763 assuaged his anger at dissident native Indians by presenting them with blankets previously used to cover people afflicted with smallpox. Without any natural immunity, smallpox ravaged the Indian population with mortality rates as high as 40 percent.[8]

At a counterterrorist council meeting, Scully reveals that the toxic organism is likely a deadly microorganism. Luckily, direct contact is needed for infection—otherwise, as Mulder states, they'd all be dead. At the Centers for Disease Control, Scully and another scientist identify the toxic biological agent as a bacterium in the genus known as *Streptococcus*. Scully is able to make this identification by using a video camera hooked up to a light microscope and seeing that the bacterial cells stay attached to one another after dividing—a hallmark of *Streptococcus*. By multiplying in this fashion, *Streptococcus*, also known as strep, produces strings of bacteria that resemble chains of linked deli hotdogs.

Scully is surprised to find that the biological warfare agent is *Streptococcus*. Strep is a poor choice for a biological weapon because,

---

[8]Amherst was rewarded for his military prowess prior to this incident by having the new town of Amherst, Massachusetts, named after him. There is even a hotel in town called the Lord Jeffrey Inn.

by not producing spores, it isn't particularly stable. Unless strep is kept moist or completely dried into a powder, it's unable to efficiently infect another organism. Scully doesn't mention the other reason why strep is an unlikely biological weapon: even the most dangerous types of strep cause few problems for most of the 20 million to 30 million Americans infected each year.

From the lesions on the dead arms merchant's face and hands, he was probably infected with one of the nastier types of strep—one of the eighty different strains known as group A strep. Group A strep can cause strep throat or, more rarely, scarlet fever and toxic shock syndrome. About fifteen thousand people a year become acquainted with group A strep after a minor skin cut. For about 1,500 of these people, the acquaintance turns deadly. Under the right conditions (which aren't yet known), group A strep can also cause a disease called necrotizing fasciitis, earning these bacteria the nickname "flesh-eating bacteria." While the bacteria don't actually eat flesh, they do secrete a dangerous toxin and cause a massive immune system overload. Blood, the body's oxygenmobile, has trouble getting to the infected tissue, causing cells at the infection site to die from oxygen deprivation. The death (necrosis) of cells is what causes the flesh to blacken and appear "eaten away." Although necrotizing fasciitis is rare, the mortality rate of 20 percent has given this disease a very high profile.

The flesh-eating disease is one of the fastest-spreading infections known; in the worst cases, death comes several hours after the first symptoms. The arms merchant, though, died within minutes after contacting the disease agent—too rapid for any normal strain of group A strep. The CDC scientist and Scully agree that the bacteria must have been deliberately genetically engineered to speed infection and enhance its deadly nature.

This is the dark side of genetic engineering—the possible generation of infectious organisms that are even more harmful than their naturally occurring cousins. Organisms can be genetically engineered in one of two ways: by tinkering with a natural gene or by adding additional genes. Some experts worry that ways will be found to insert venom-producing genes from cobras or scorpions into bacteria, generating organisms that kill better and faster. Genetically engineering organisms so that they aren't recognized by a vaccinated person's immune system is another potentially deadly use of this science, and the one chosen by the terrorists in "The Pine Bluff Variant."

The CDC scientist tells Scully that the lethal strain of strep that killed the arms merchant has been deliberately genetically engineered so that each bacterium is surrounded by a capsule, what he calls a synthetic protectant coat. Bacteria that are surrounded by capsules are much less visible to the immune system. The most deadly bacterium known—*Bacillus anthracis*—the causal agent of anthrax disease, contains a capsule made out of chains of a single amino acid. The particular strain of strep that causes pneumonia (*Streptococcus pneumoniae*) is naturally surrounded by a capsule made out of sugars. When *Streptococcus pneumoniae* has a mutation in one of the fifteen genes needed to make the sugar capsule, the bacteria are no longer covered with a capsule and no longer cause disease. The capsule is therefore providing a shield that insulates the bacteria from prowling immune system macrophages—the cells that eat invading microorganisms.[9] If the strep strain used by the terrorists was genetically engineered to produce such a capsule, then it could very easily acquire additional infectious properties by becoming less visible to the immune system.

Scully is concerned about how the arms merchant became infected with the bacterium. Most developers of biological weapons try to find ways to disseminate such agents through the air, using some type of aerosol spray. A major stumbling block in the use of biological weapons is the delivery system. Inhaling a single bacterium or having a single group A strep cell enter a wound will not cause an infection. For the deadly anthrax bacterium, about eight thousand to ten thousand spores need to be inhaled before half of the infected people will die.

---

[9]*Streptococcus pneumoniae* and its capsule played an important role in the history of molecular biology. In 1928, English physician Frederick Griffith found that bacteria making a capsule can transfer "something" to bacteria that don't make a capsule and convert them into ones that now make a capsule. What gets transferred was later identified as DNA—specifically the DNA genes that specify the making of the capsule. This was the first inkling that DNA was in fact the genetic material of cells. It was also the first indication that bacteria are able to transfer DNA between themselves, which happens whenever bacteria "mate."

The terrorists in "The Pine Bluff Variant" decide that their delivery system for the deadly strep agent needs some additional testing. The test site chosen is a movie theater in Ohio. A terrorist arrives at the movie theater holding an unlabeled aerosol spray can. By the time anyone notices that the patrons are actually staying in their seats through the end credits, fourteen moviegoers and employees are dead. Scully determines that the victims have the same symptoms as the arms merchant, their skin appearing horribly blistered and blackened. No traces of unusual organisms are found in the air, suggesting that the theater's recirculating air supply wasn't used to disseminate the bacteria. Scully becomes convinced that the terrorists are impregnating something that all the patrons touched but can't find anything on an obvious item, like ticket stubs.

During the heyday of the American biological warfare program (1943 to 1966), testing was also conducted on how to spread biological warfare agents. While the United States government was more mindful of harming innocent civilians than "The Pine Bluff Variant" terrorists (although Mulder might disagree), the tests that were conducted are still very controversial today. Laboratory workers posing as travelers walked through Washington National Airport discreetly spraying nontoxic *Bacillus subtilis* into the air while monitoring stations recorded how quickly the little bacteria traveled; using a light bulb container that was dropped onto New York subway tracks, only twenty minutes was needed for the released bacteria to spread throughout the commuter-crowded system; a navy ship docked in San Francisco Bay sprayed the bacterium *Serratia marcescens* into the air, which was soon picked up by monitoring stations more than thirty miles away. Eleven San Franciscans developed rare infections from this bacterium within a few days of the test and one man died. Even Scully would have had trouble buying the "It was a coincidence" story that the government released during the trial to compensate the victim's family. In all, 239 open-air biological warfare tests were conducted within populated areas of the United States using bacteria that were considered harmless but, in rare cases, were also known to cause illness or death.

When Scully sees Mulder driving off with a terrorist, she follows but is soon cornered by other federal agents. She learns from her boss, Assistant Director Skinner, that Mulder is involved in a dangerous covert operation to smoke out the terrorists and discover their ultimate aim. Scully tells Skinner that the biological weapon being used by the terrorists was probably developed by the CIA. Skinner maintains that it's more likely of Russian origin. If this episode were based on a real-life incident, Skinner would probably be correct. America's very active offensive biological warfare program, centered at Fort Detrick in Maryland, was disbanded in 1969 under the directive of President Richard Nixon and all toxic organisms destroyed. Nixon believed that America's substantial nuclear weapon arsenal could act as a deterrent in the event of a biological warfare attack by another country. He also believed that others—not necessarily friendly others—would benefit from America's extensive research on the production, dissemination, and storage of biological agents more than would America. It isn't known if the abhorrent nature of the weapons, which had just been tested on animals aboard ships in the Pacific Ocean, also influenced Nixon's unilateral decision.

The former Soviet Union, the U.S., and 138 other governments signed an agreement in 1972 not to develop, produce, stockpile, or otherwise acquire or retain microbial or other biological agents or toxins, whatever their origin or method of production. The Soviet Union apparently interpreted the agreement to mean that they could develop, produce, and stockpile twenty tons of plague bacteria, twenty tons of smallpox virus, and hundreds of tons of anthrax bacteria over the following twenty years. At the height of the program, seventy thousand Soviet scientists and technicians were secretly involved in developing biological weapons of mass destruction.

Western countries might still be in the dark about the Soviets' biological warfare program if not for a little news blip in 1979 from a city of 1.2 million people called Sverdlovsk. It began with the death of a goat in the Chkalovsky neighborhood. Soon adults living near the goat became sick. Then people all over Chkalovsky started complaining of pain so intense that some were misdiagnosed as having heart attacks. Pneumonia was also suspected. Death came within hours or days of the first symptoms.

People, all adults and mostly men, started dying on the streets, at work, and at home. Doctors from Moscow arrived on the scene and correctly diagnosed the problem: anthrax. Although anthrax is mainly a disease of animals like cows, sheep, and goats, humans can get anthrax by touching infected animals, but such infections are rarely lethal. Eating or inhaling anthrax spores is much more lethal. The Soviet Union provided an explanation for the sixty-six mysterious deaths: people ate contaminated beef purchased from a private butcher. In this short statement, Soviet officials provided a reasonable explanation plus a little lesson for shopping at state-run butcher shops. The explanation was accepted by many Western scientists, in particular, the highly respected Matthew Meselson of Harvard University. Disbelievers were mainly officials in the American government, who pointed to a likely mishap at a biological warfare facility suspected of being located at the army base in the city of Sverdlovsk.

Although Meselson thought that the Soviets' contaminated beef explanation was plausible, he nonetheless wanted to investigate the incident firsthand. In 1992, Meselson and a group of American and Russian scientists visited Sverdlovsk and began to uncover the mystery of the anthrax epidemic. They found that the Soviet secret police, the KGB, had confiscated all the public health records and bodies of the victims. Finding a list of patients who died, Meselson and his team were able to precisely map where each of the victims lived and worked. The geographical distribution was astonishing. Nearly everyone lived or worked in a very narrow corridor extending for about two and one half miles. The corridor was so narrow that in one large factory, only workers at one end died. At the northern tip of the zone was—big surprise—a military microbiology facility.

In 1992, Russian President Boris Yeltsin admitted that the anthrax deaths were due to a mishap at the army base. According to Dr. Kanatjan Alibekov, the director of one of the main Soviet biological warfare factories, the Sverdlovsk facility was responsible for the continuous manufacturing of anthrax biological weapons. The mishap at the factory was a tiny one. Of the hundreds of tons of anthrax being processed and stored, less than four ounces were released when a filter was inadvertently left off an air vent five days before the first deaths. Recent examination of the remains of some of the victims (stored without knowledge of the government by doctors in Sverdlovsk) indicates that the victims were exposed to at least four different *Bacillus anthracis* strains. The interpretation of these results? The Soviets were actively seeking ways to render the current anthrax vaccine impotent. According to U.S. government officials, they weren't successful.

Governments aren't the only ones with biological warfare programs. The threat posed by domestic terrorist groups leaped from conjecture to reality in 1995. In a crowded subway beneath the streets of Tokyo, the Japanese religious Aum Shinrikyo sect released the nerve gas sarin, killing a dozen people and causing the hospitalization of many thousands of commuters. When the Aum compound was raided, police found that chemical weapons were only part of the arsenal. The cult was actively working on genetically inserting the gene for botulism toxin into the common bacterium *E. coli* and devices to spray the bacteria over a large population. The Aum sect was also trying to expand their repertoire of deadly pathogens. In 1992, they sent a team to Zaire to assist in the treatment of Ebola victims with the hope of acquiring a sample of the virus. They were apparently unsuccessful.

Odorless, tasteless, easy to produce, and very cheap—it is no wonder that so many experts on terrorism are concerned about the use of biological warfare agents by ethically deficient governments and malcontents around the world. For these reasons, Mulder's infiltration of a domestic terrorist organization bent on using such weapons is so important. After their successful test in the Ohio movie theater, the terrorists are ready to release the bacteria on a much larger segment of the population by using the Federal Reserve to disperse tainted money. Still unaware of the deadly nature of the plot, Mulder provides the group with Federal Reserve bank pickup information. The terrorist group then robs one of the banks on the list. The robbery, though, is a ruse—the real objective is to spray the bacteria on money that is ready for pickup. Their plans are foiled when Scully realizes that the target is a bank, and recognizes Mulder on the video monitor of one bank that was recently robbed. By the time Mulder gets free of the terrorists, he is relieved to find Scully and decontamination crews hard at work at the bank but upset to find the government denying that anything took place.

While the reasons for denying the incident are highly suspicious to Mulder, concern about panicking the public probably played at least a minor role in trying to whitewash the incident. Just how panicky the public can get was seen in a well-publicized incident that occurred in Las Vegas in early 1998. The FBI were tipped off that two men claimed to have a large quantity of "military grade" *Bacillus anthracis* in their car. Stores in Las Vegas quickly ran out of gas masks as news of a possible source of anthrax spread. The bacteria turned out to be missing their immune system–shielding capsules, and thus usable only for vaccine purposes (without the capsule the bacteria are unable to multiply in the body long enough to cause harm). One of the men arrested was well known to the FBI. In 1995, Larry Wayne Harris paid $240 to a supply house for several vials of bubonic plague bacteria, which he kept in the glove compartment of his car. Despite ties to a violent white supremacist organization and a background in microbiology, he could only be convicted of mail and wire fraud. Given these disturbing incidents, one can only hope that real-life equivalents of Mulder and Scully are busy trying to ferret out domestic terrorists who would engage in this ultimate insult to nature.

We Are All Frogs

EXT. MUDDY SHORELINE–LATE AFTERNOON

FARRADAY

A frog holocaust is currently being executed, Dr. Bailey. And man is the executioner.

Bailey has heard enough from this tree-hugging liberal.

BAILEY

You're a biologist, Farraday. You've never heard of survival of the fittest?

With that, Bailey turns and walks away—Farraday calling after him:

FARRADAY

Just remember, that rule also applies to mankind!

But Bailey just continues on . . .

FARRADAY

You can't turn your back on nature, or nature will turn its back on you!

—"Quagmire"

Where have all the frogs gone? This is the mystery confronting naturalist Dr. Paul Farraday as he studies amphibians in the Blue Ridge Mountains of Georgia in the *X-Files* episode "Quagmire." Frogs, which used to thrive in and around the Heuvelmans Lake area, now barely register a ripple on the frog Richter scale. Farraday is particularly concerned about *Rana sphenocephala,* the southern leopard frog, once so ubiquitous from southern New York State to Florida but now nearly just a memory. Farraday tries to convince an unimpressed Dr. Bailey of the United States Forestry Service that protective measures are necessary before all the frogs disappear. Dr. Bailey replies that frogs are declining all over the world, not just at Heuvelmans Lake, and this is merely a case of survival of the fittest. After leaving the company of Dr. Farraday, Dr. Bailey learns the true meaning of nature taking its course when he finds himself on the menu of a creature far more fit to survive in the wilderness than fragile Forestry Service employees.

As the late Dr. Bailey mentioned, amphibians, those diminutive frogs, toads, salamanders, and newts, have not been faring well in the modern world. In the past fifty years, many species have diminished to the point where they are uncommon, endangered, or even extinct. Before the late 1980s, the reasons for the downward trend in populations of amphibians was reasonably obvious. Amphibians, which must lay eggs in water and keep themselves moist, need the kind of wetland areas that have been declining in civilized areas. The continental U.S. has destroyed over half of their original 221 million acres of wetlands. Consider, for example, the natural ponds that used to dot the desert landscape of southern Nevada and were once home to the Las Vegas frog. In their place are highways, hotels, and golf courses, and the Las Vegas frog is forever gone. In Arizona, stocking lakes with exotic species like bass and sunfish make for happy local fisherman but unhappy local amphibians, who become food for the fish that have invaded their homes. Even tidying up unsightly mud holes that form in the spring or after summer rainstorms leads to deadly eviction notices for the amphibians that need these habitats to survive. Pollution of streams and rivers, UV radiation from loss of the ozone layer, acid rain—all are contributing toward slowly diminishing the numbers of amphibians.

Still, amphibians are very hardy creatures. In their modern forms, they have been around for over 200 million years. The problem of amphibian decline, while real, wasn't causing many outside environmental circles to take notice. Since amphibians managed to survive the catastrophic event that killed the dinosaurs, even naturalists like Dr. Farraday, if queried ten years ago, would probably have agreed that enough pristine areas of the planet remained for most amphibian species to long outlive the much younger human race.

What concerns the present-day Dr. Farradays of the world is that the slow amphibian decline has turned into a sharp plummet in the past ten years. It doesn't take a trained biologist to know that something ominous is afoot. Most everyone who lives near a pond or lake appreciates that something is happening to the water's inhabitants. In 1988, the cacophony of nightly sounds emanating from the frogs and toads that resided in the natural pond that borders my house was so raucous that sleeping was problematic on hot summer days. Yet for the past five years or so, there has been scarcely a whisper at night—sometimes just the cry of a single bullfrog. Farther west from Massachusetts, in the rugged beauty of Yosemite National Park, 50 to 80 percent of the frogs have vanished. In Australia, fourteen frog species have seen their population sizes suddenly crash, with two becoming extinct. In Panama and Costa Rica, amphibians that once thrived are now gone forever. What began as a gradual amphibian decline in highly populated areas in the latter half of the twentieth century has turned into a plummeting spiral in the amphibian population worldwide; as Dr. Farraday puts it, a frog holocaust is under way. In 1998, estimates are that one thousand of the four thousand species of amphibians in the world are vulnerable or in immediate danger of extinction. This is a problem that can no longer be ignored and is the reason why biologists like Dr. Farraday are now actively monitoring frog populations worldwide.

Dr. Farraday is trying to have various frogs placed on the list of endangered species. The Endangered Species Act, passed in 1973, was the most sweeping environmental legislation ever made into law. Its goal was to protect plants and animals that were threatened because of human endeavors. It signified a hope that humans and animals could coexist—that there was a better choice besides plundering the land for its wealth irrespective of the consequences to tiny native inhabitants, and living in a dark cave sucking on raw vegetables. People chose to try and save species that could not save themselves while still allowing land to be developed and used. With the passage of the act, no longer could endangered species teetering on the verge of extinction or species threatened with becoming endangered be sold or killed with impunity. If a parcel of land was home to some species on the list, it could no longer be developed without some consideration for its fragile inhabitants.

Any citizen of the U.S. can petition the Fish and Wildlife Service to place a species on the list. Like Dr. Farraday's petition, it must be backed up with solid research and usually document some adverse activity that is having or will have a major impact on a dwindling population. Like all petitions, Dr. Farraday's proposal for the southern leopard frog must have been published by the government in the Federal Register. After a period of time for the public to comment, Dr. Bailey was sent out to further assess the situation before the Fish and Wildlife Service could approve or disapprove the petition. Dr. Bailey was being realistic when he told an angry Dr. Farraday that not all frogs could be put on the list. What Bailey was referring to was the substantial expense that must be set aside for all species on the list. For not only are such species protected from being bought and sold, but a recovery plan must be drawn up to ensure that the populations increase in size. Species recovery is not cheap. Land must be bought and set aside for habitat, captive breeding programs initiated, population sizes monitored, and the public educated. Funds can easily exceed several million dollars for recovery of a single species.

While the act has many detractors, primarily developers who have other plans for land occupied by an endangered species, there are many success stories. Nearly fifty-one of the ninety-one birds that were threatened or endangered are now either stable or increasing in number. Bald eagles, peregrine falcons, and brown pelicans all owe their recent recoveries to the Endangered Species Act. Of the 128 original species placed on the list in 1973, almost 60 percent have either recovered, improved, or are at least stabilized. Yet as Dr. Bailey pointed out, not all endangered species will be placed on the list. While over 1,450 are presently listed, more than four thousand species are waiting for their final evaluation—the yea or nay that decides who will survive into the twenty-first century and who will not.

The Act does have other critics who believe that it has had an unintended consequence. By focusing such heroic efforts on species only when they are on the brink of extinction, scarce resources are diverted from a far larger number of species habitats that might be nurtured while they are still somewhat robust. It's like spending huge sums only on patients with life-threatening diseases, with nothing left over for prevention.

Amphibians are not the only creatures missing from Heuvelmans Lake. A Boy Scout troop leader also disappeared from the same site as the missing and presumed eaten Dr. Bailey. Since Bailey was a federal employee, the FBI is called in to investigate. At first, Scully is confused as to why Mulder wanted assignment to the case; then she sees a sign advertising Mulder's prime suspect in the disappearance of Dr. Bailey, *Big Blue the Southern Serpent. Spot him at Heuvelmans Lake.* Once at the lake, Dr. Farraday tells Mulder and Scully that Big Blue and other monstrous lake creatures are pseudoscientific nonsense. Dr. Farraday probably bases his belief on the complete lack of remains recovered for any mythical lake creatures—and a large number of creatures would be necessary to keep a breeding population alive without suffering the ill consequences of inbreeding. Lakes where dinosaurlike creatures have been "sighted," like Loch Ness in Scotland and Lake Champlain in the U.S., are also the homes of monstrous hidden water waves called seiches, which are powerful enough to spew up debris that could be mistaken for monsters.

True to his nature, Mulder prefers to believe that Big Blue is a throwback or, more likely, a descendant of prehistoric creatures thought to be extinct but still very much alive. If so, Big Blue wouldn't be the first prehistoric creature to be removed from the extinct species list.

In 1938, Marjorie Latimer, the curator of a local museum, glanced over the catch of some local fishermen at a port northeast of Cape Town, South Africa, and couldn't believe her eyes. Peeking out from beneath a pile of rays and sharks was a giant, five-feet-long, pale blue fish marked with iridescent silver. A sketch sent to L.L.B. Smith, a Rhodes University professor, resulted in a hurried cable back to Latimer to preserve the fish's skeleton, which unfortunately had been discarded during mounting. When Smith came to view the fish, he immediately identified it as a coelacanth. What was so remarkable was that coelacanths had been extinct for about 80 million years—or so it was thought. The prehistoric fish swiftly became known as the "most important zoological find of the century," a title it continues to hold to this day (unless, of course, the waning days of the century provide proof that the Martian meteorite fossils are truly life from Mars).

If Big Blue turns out to be a living plesiosaur, the event couldn't attract more attention than that paid when finding the living coelacanth. The reason for all the excitement had to do with what coelacanths represent in the fossil record. Coelacanths first appeared about 360 million years ago during the dawn of the first land animals—fish that wriggled their way out of the sea. Many scientists were convinced that the coelacanth represented a link between those adventurous fish and their land-animal descendants. Relying only on the fossil record, this view was widely accepted since the back fin bones of the coelacanth were jointed, like those found in arms and legs.

What the South African fishermen had dragged into their nets was a living fossil, but without the skeleton, it was difficult to convince others conclusively that modern coelacanth and fossilized skeletons of ancient coelacanth were the same. The race was on to find another coelacanth. For fourteen years, reward posters and word of mouth failed to turn up a second coelacanth. Finally, in 1952, fishermen from the Comoro Islands hooked a coelacanth in the presence of someone who was familiar with the wanted posters. Dinner was sacrificed for the benefit of science.[10] As it turned out, the fishermen told scientists that they were familiar with this fish and frequently caught it between December and March (the population in the Comoro Islands is estimated to be about five hundred). In addition to the coelacanth living near the Comoro Islands, another population was recently found off Indonesia that are brown instead of

---

[10]Not that it was much of a sacrifice—the taste of coelacanth has been described to me by one who knows as "truly disgusting."

blue. Despite considerable attempts, all efforts to capture a live coelacanth have failed.

Still, the ability to study even a dead living fossil has proven immensely exciting for scientists with a passion for fish. I will always remember the scene about ten years ago when my fish-besotted
colleague Dr. Willie Bemis ran down the halls of the Morrill Science
Building screaming, "The coelacanth is here! The coelacanth is here!"
As I rushed from my office, I spied this huge, ugly, smelly fish lying
on a large cart. Much to the relief of noses throughout the building,
Willie soon wheeled his prize into his laboratory for a thorough examination. Over the next few years, Willie delighted in studying this
relic with its collection of unique attributes, including a previously
unknown sense organ that may be related to structures that allow
hearing in air, and a wall through its braincase that separates the
nasal organs and eye from the ear and brain. The coelacanth's physiology, though, revealed little about the evolution of land animals. As
Willie explained to me after his examination of the fish was completed, it's more romantic than real to consider the coelacanth as an
icon for the transition between life in the sea and life on land (that title is currently held by the lungfish). Finding living coelacanth, however, did spark a huge amount of interest in the evolution of fish and
land animals. Willie's coelacanth can now be seen by everyone at its
new home in the American Museum of Natural History in Washington, D.C.

Although freshwater coelacanth might still exist and coelacanth
mouths have very sharp teeth, no coelacanth was responsible for the
deaths at Heuvelmans Lake. The existence of some other hungry prehistoric creature remains a possibility in the mind of Mulder when
the half-eaten Boy Scout troop leader is recovered and four more local residents die. With so many deaths and the recent loss of the frogs
from the lake, Mulder begins to wonder if there is a link between
frog decline and Big Blue's new appetite. If Big Blue had eaten all the
frogs, maybe he needed to switch to a different, more abundant food
supply.

Mulder's speculation that the frogs vanished because they were
all eaten by some enormous sea creature might explain the amphibian decline at Heuvelmans Lake but it doesn't explain the decline in
amphibians worldwide. The reason for the downward spiral in frog

populations from pristine locations like Panama, Costa Rica, and the Australian rain forest has only recently been discovered. Not surprisingly, the culprits aren't prehistoric sea monsters. Scientists from Australia, Great Britain, and the U.S. determined that the dead and dying frogs suffer from a massive skin infection. Since frogs drink and breathe through their skin, the infections are causing the frogs to suffocate and die.

The frog killer is an organism right out of the *X-Files* episode "El Mundo Gira." However, instead of an alien enzyme causing ordinary fungi to become deadly pathogens, something much closer to home turned a normally harmless fungus into a killer. The fungus, a type of chytrid fungus, is found in many streamside locations throughout the world. Like all fungi, they help to decompose dead organisms. Chytrid fungi were previously know to infect some simple organisms but nothing as complex as an amphibian. Somehow these most ancient of all the fungi—dating back some 600 million years—have suddenly discovered a scavenger's dream: a way to supply their own dead meals.

What disturbs many biologists is the question, why now? What has suddenly happened to make frogs vulnerable to a fungus with which they have likely coexisted for millions of years? The possibilities are every bit as ominous as alien enzymes from space. It could be global warming causing a rise in temperatures; or higher UV levels from the declining ozone layer; or maybe it's pesticides, applied to crops thousands of miles away, that have made their way into the most isolated of environments. What is happening to frogs worldwide is probably affecting far, far more than just frogs. Frogs are just a single population in a community of populations. Whatever has made amphibians more susceptible to disease is also affecting everything else, including us. Frogs may just be an early warning sign that something is going very wrong.

Whether or not the frogs at Heuvelmans Lake died off from being eaten by a large predator or from the chytrid fungus is not known. What is known is that the death of the frogs has upset the food chain in the area; whatever was eating the frogs is now eating people. When a species is lost from a local ecosystem, the reverberations can echo up and down the food chain. Whatever was eating the frogs must now search for new food, and whatever the frogs were eating may now overwhelm the habitat. This scenario plays out every time a species goes extinct. It's also the situation facing the Australians as they attempt to deal with their rabbit problem. If the calcivirus designed to kill off the rabbits is successful, what will the predators turn to when there are no more rabbits to eat?

When the lake creature makes a grab for Dr. Farraday, who is carrying a bag of frogs bred in captivity that he is returning to the wild, it becomes clear that frogs are the preferred food. When the creature decides to eat Mulder for dessert, Mulder shoots it repeatedly, and can't help but be disappointed by what lies dead at his feet. Big Blue turns out to be more aptly called "big grayish-black"; not a plesiosaur, but a huge alligator, rivaling the record length of nineteen feet two inches. If Mulder was simply looking for a creature from the time of the dinosaurs, he did find one. Alligators have existed virtually unchanged for 65 million years. What made this gator switch from more typical meals of fish, turtles, snakes, small mammals, sticks, stones, fishing lures, and aluminum cans to humans is not known. Perhaps, like so many freshwater lakes and rivers, the local animal population at Heuvelmans Lake was displaced by the stocking of a few exotic species of fish not palatable to the alligator.

Fortunately, Dr. Farraday survives the alligator attack and can continue with his work to have *Rana sphenocephala* placed on the endangered species list. Although the late Dr. Bailey implied that Dr. Farraday's research was not sufficiently conclusive for the frog to be listed, politics and not science may have dictated the outcome. Consider the northern spotted owl. Pressure from groups with commercial interests in the owl's natural habitat killed the original petition. Perhaps Dr. Farraday will use the owls as an example and sue the Forestry and Wildlife Service for inclusion of his frog on the list. And just like the current protection of the owl's old-growth trees is helping all inhabitants of some Northwestern U.S. forests, protecting the frogs' habitat will help other—possibly many other—identified and unidentified species survive the beginning of at least one more millennium.

Afterword

The life of a research scientist is filled with mysteries as complex as any that appear on *The X-Files*. We are Scullys: constantly questioning and exploring, formulating hypotheses, and dismissing them when supporting evidence isn't found. Much of the knowledge regarding organisms that share our planet comes from so-called "small science"—individual investigators at universities, colleges, and museums who along with their students are deciphering such mysteries as how life began and branched into millions of current species; how cells operate and communicate within complex organisms; how genes provide the blueprints for appearance and behavior; and how life deals with being surrounded by potential pathogens. As the complex layers of life are minutely peeled back, it is increasingly evident that every organism has secrets to share that help us better understand the human animal. There is so much left to learn.

The next few decades should prove immensely exciting. Just around the corner are replacement organs grown from a person's own cells, plants genetically engineered to produce vaccines and thrive without fertilizers and insecticides, treatment of diseases from the inside out with gene therapy, the tantalizing possibilities of nanotechnology, and, maybe, the discovery that Earth doesn't have a monopoly on life in the solar system. If we can clean up our ecological mistakes and avoid new ones, and control population levels so that diminishing resources are not overexploited, then the future need not be fought but rather welcomed with open arms.

The scientists who will achieve these advancements will come from today's children: children who will choose science over vastly more lucrative and less stressful careers. Koshi Dhingra, a graduate student at Columbia University, recently surveyed a large number of ninth-grade students for where they were exposed to science on television. The top answers were PBS documentaries, the Discovery Channel, news programs and . . . *The X-Files*. The students told her that they had selected *The X-Files* because of the realism with which Scully uses science; the accurate representations of jobs dealing with science; that science is used to disprove theories on aliens; and that science is used to make the supernatural seem more believable even though they understand that the supernatural events aren't realistic. Critics who claim *The X-Files* is harmful to the public's awareness of science would probably be amazed to learn how many students in my freshman biology class point to the favorable portrayal of science and scientists on *The X-Files* as one reason for their interest in science.

When *The X-Files* ends its highly successful run, I will be among the many fans who will have to find another way to spend Sunday evenings. But in addition, I will miss those late-night phone calls from Chris Carter inquiring how people can suddenly break out in reptilian scales, and oh, by the way, it needs to involve something they ate. Life just won't be the same.

Acknowledgments

This book could not have been possible without the help of a large number of people. My thanks to student researchers Sian Gramates, Shelley Schlief, Johanna Rodrigues, John Bohannon, Connie Villalba, David Klein, and Seth Eichenlaub at the University of Massachusetts as well as Koshi Dhingra from Columbia University for sharing the results of her study on where students learn science. I am indebted to University of Massachusetts colleagues Drs. Sue Leschine, Jim Robl, Guy Lanza, Sandy Petersen, Tom Zoeller, Derek Lovley, Anne Averill, Willie Bemis, Ed Klekowski, and Judit Pogany for stimulating conversations, reviewing various chapters, and help with reference material. Much thanks to scientists from other institutions, Drs. T. C. Onstott, Roy Gallant, Geoffrey Briggs, and George Martin for responding so positively to my requests for assistance. Keeping me from complicating simple stories with scientific jargon were my nonscientist readers Dori Pierson Carter, Sondra Simon, and Mayo Simon. And finally, special thanks to Chris Carter for producing a show of such quality; Bill Rosen, my wonderful editor at Simon and Schuster; Esmond Harmsworth, my agent at Zachary Shuster, for the book writing suggestion; the members of my laboratory for putting up with a year of so many distractions; and the National Science Foundation for supporting so much fascinating research, including my own work on viruses, and for their tremendous efforts in science education.

Index